Jim Miller

EXPERIMENTS IN METEOROLOGY

Investigations for the Amateur Scientist

Leslie W. Trowbridge

Doubleday & Company, Inc.

Garden City, New York

1973, 270 p.

ACKNOWLEDGMENTS

I wish to acknowledge the assistance of my three sons, David, Thomas, and Howard Trowbridge, whose ideas for simple meteorological experiments have contributed measurably in the preparation of this book.

LESLIE W. TROWBRIDGE

ISBN: 0-385-08238-x
Library of Congress Catalog Card Number 70–171324

Printed in the United States of America

SOURCE MATERIALS

INVESTIGATION 9. Adapted from an unpublished report by Donald T. Acheson, U. S. Weather Bureau.

INVESTIGATION 11. Suggested by Mr. F. I. Badgley.

INVESTIGATION 13. Photographing Snow Crystals

INVESTIGATION 14. Adapted from James E. Juisto and Roland J. Pilie, "Condensation Nuclei Experiments with Simple Apparatus," *Weatherwise,* December 1958, pp. 206–8.

INVESTIGATION 15. Adapted from David Fultz and Robert Kaylor, "Dishpan Hurricanes Help Study of Large Storms," *Science Newsletter,* Vol. 71:201, 1957; David Fultz, "A Survey of Certain Thermally and Mechanically Driven Systems of Meteorological Interest, Fluid Models in Geophysics," *Proceedings of First Symposium on the Use of Models in Geophysical Fluid Dynamics,* American Meteorological Society; and Frank R. Bellaire and Albert W. Stohrer, *The Development of Laboratory and Demonstration Equipment for Meteorological Instruction,* University of Michigan, 1963.

INVESTIGATION 17. Adapted from James E. Miller, "A Tornado Model and the Fire Whirlwind," *Weatherwise,* Vol. 8, No. 3, p. 88; and Bellaire and Stohrer, cited above.

INVESTIGATION 20. Adapted from *Instructions for Making Pilot Balloon Observations,* W. B. 1278 Circular O, Aerological Division, U. S. Government Printing Office, Washington, D.C.

INVESTIGATION 21. Adapted from Charles A. Laird, "How to Construct a Density Channel," *Weatherwise,* August 1948, p. 87.

INVESTIGATION 23. Adapted from H. H. Lettau and J. K. Sparkman, "Hot Plate Mirages," unpublished paper, Department of Meteorology, University of Wisconsin.

INVESTIGATION 25. Adapted from Hans Neuberger and George Nicholas, *Manual of Lecture Demonstrations, Laboratory Experiments, and Observational Equipment for Teaching Elementary Meteorology in Schools and Colleges,* Pennsylvania State University, 1962, p. 127.

CONTENTS

EXPERIMENTS IN METEOROLOGY

Chapter 1

INTRODUCTION

A jet plane is flying from Portland, Oregon, to Chicago, cruising at 31,000 feet through clear skies. As it passes over the mountains in western Wyoming, there is a brief but ominous buffeting, which causes the whole plane to vibrate. Suddenly, the plane drops, a terrifying plunge as fast as the free fall of gravity. Loose objects rise rapidly to the ceiling. The plunge lasts less than a minute, but the plane falls 3,000 feet before resuming its normal flight—normal, that is, except for a much shaken-up crew and unnerved passengers.

What has this to do with meteorology? What caused the precipitous drop? Could the region of turbulence have been predicted, and thereby avoided? As yet, meteorology doesn't have all the answers to these questions. Clear air turbulence—"CAT"—is what the plane encountered. But its cause is not yet fully understood.

A puffy, white cumulus cloud begins to grow larger in the afternoon. Its top swells and billows, and finally starts to appear feathery and wispy. High-altitude winds carry the wispy portion ahead of the main part of the cumulonimbus cloud, giving it the appearance of an anvil. Suddenly, a brilliant flash of light is followed almost immediately by a rending crash of thunder, and rain begins to cascade from the cloud.

Which came first—the lightning or the rain? What tears charge from charge? Do splitting raindrops cause lightning—or do electric charges in the air cause the raindrops to form?

These are more questions for which meteorology still has uncertain answers.

The science of meteorology is still young. Only recently has it begun to pass from the stage of wonder, intuition, and speculation, to a science of observation, measurement, and prediction. Many significant advances have been made in the past fifty years. Observation techniques have improved from entirely ground-based records to sophisticated measurements at all levels of the atmosphere up to 100 miles or more. Even more accurate information is needed, as industries and government agencies clamor for better weather information.

But what can the meteorologist really say about the vast atmosphere? Can he keep track of even one of the myriads of molecules that make up the sea of air? Can he predict what will happen to the infinite swirls and whirlpools of air that so greatly affect the inhabitants of the earth?

This is a book to stimulate thinking—about the unsolved problems in meteorology. In most cases, the equipment suggested to conduct the experiments and observations described here can be made from simple materials, easily and inexpensively obtained. A few more advanced studies call for an investment in radio parts, camera equipment, or, in one case, a miniature rocket; in no case does the required budget for equipment exceed $100. Wherever possible, actual data obtained in trial investigations have been included to assist the amateur investigator in setting up his own experiments. But, as you proceed, you will undoubtedly start modifying and adapting, and perhaps even inventing brand-new ways of studying these problems.

Much of what we experience as weather can be attributed to water in the atmosphere and the particular form, or combination of forms, in which it appears. Because the weather is always with us—and there are ample opportunities to watch clouds form, or to feel the changing levels of humidity, temperature, and pressure in the atmosphere—this is an area of study that is easily accessible to the amateur investigator. Many puzzling questions about cloud formation and precipitation remain unanswered, and the challenges are great. Two chapters on clouds and atmospheric moisture will start you on your way.

We are also, in our daily activities, aware of air movement and storms. The natural forces and conditions that cause air movement—radiation from the sun, the force of gravity, and the earth's rotation—constantly exert their combined effects. With care and persistence, these effects can be studied. As Chapter 4 will describe, models that simulate certain natural conditions can be most helpful. In other cases, measurements taken with suitable instruments can provide data to be analyzed for a better understanding of possible cause-and-effect relationships.

In the early days of meteorology, much was learned by studying the optical effects produced by varying conditions in the atmosphere. Such study is still a good source of information. Knowledge of the principles of ray optics, that is, what light does when it is transmitted, reflected, refracted, dispersed, diffracted, and absorbed, is essential to careful analysis of the atmosphere by optical methods. Fortunately, light can be depended upon to perform in predictable ways, and the problems here are mostly concerned with accurate observations and interpretation. Chapter 5 proposes several ways to "see" the atmosphere. Rainbows, halos, and mirages are commonplace, and the sky is constantly changing its appearance. All these events have meteorological significance.

The more awesome and violent displays of atmospheric phenomena frequently have to do with electrical effects. Being transitory, these effects are not easy to study. How would you obtain photographs of lightning, for example? What can you learn about the normal distribution of electrical charge in the atmosphere in fair-weather conditions? How does it change during stormy weather? Chapter 6 on atmospheric electricity will suggest some possible approaches to the solution of these problems.

The final chapter describes some methods of studying atmospheric structure and composition, including two investigations of hard-to-get-at sections of the atmosphere: air immediately above the ground, the study of which is called micrometeorology, and air a thousand feet or more above the earth, which must be reached by balloons or rockets.

Appendices gives practical information on how and where to obtain supplies, and a bibliography suggests further reading.

Much of our present knowledge was gained by amateurs like yourself. Although new secrets of nature may not yield to your investigations, these experiments will serve as groundwork to help you understand basic concepts, and, as you learn more, perhaps you will play a part in scientific discovery. What it takes is interest, and inventiveness, and somewhat greater than average persistence! These, plus good luck, will often bring unexpected rewards.

NOTE: The experiments included in this volume have been marked to show relative difficulty involved in carrying them out. Those marked E are relatively easy; the M experiments are a little more difficult; those marked D are the most difficult or will take a little more time to carry out. However, all of the projects will be workable by most readers.

Chapter 2

CLOUD STUDIES

One of the most exciting areas in meteorology is the study of clouds and precipitation. And, as we mentioned earlier, this is an area of investigation about which much can be learned through relatively simple demonstrations, measurements, and observations. What causes clouds? How high are they, and how fast do they move? What kinds of clouds are there? How much of the sky is usually covered by clouds? These and other questions are the subject of our investigations in this chapter.

The meteorologist is interested in clouds for many reasons, not least of which are their beauty and uniqueness in the known universe. No other planet or satellite is known to have an identical cover of water droplets—for that is what a cloud is. Photographs from distant space taken by astronauts have shown strikingly beautiful cloud systems covering nearly half the earth's surface. Can there be any doubt that this unique mantle must play an important role in the earth's weather and on the activities of its inhabitants?

Consider first how clouds affect sunlight that reaches the earth. Does sunlight go through clouds? Generally, not very far. Of course, clouds are not completely opaque, but they succeed in blocking a large fraction of the sunlight striking them from space. The term "albedo" refers to the reflectivity of a surface. The average albedo of the earth is about 34 per cent, which means that, on the average, 34 per cent of the incoming sunlight is reflected back out to space. Much of this reflection takes place from cloud tops. If you have flown above the clouds, you have probably been impressed by the brilliant whiteness of the cloud carpet below you. This whiteness is indicative of high reflectivity.

But clouds also absorb radiation from below. The heated earth sends out long-wavelength radiation, which strikes the clouds and is trapped. Some of this trapped radiation is reradiated downward and absorbed again by the earth, with a resulting increase in the temperature of the earth's surface and the lower layers of the atmosphere. Because this activity resembles what takes place in a glass-roofed greenhouse, it is called the "greenhouse effect." Clouds, then, help to maintain a higher temperature in the lower parts of the atmosphere. The meteorologist makes use of this information when forecasting temperatures. If the sky is likely to remain cloudy for some time, he will probably predict smaller drops in temperature at night than if the sky is clear.

Another effect of much concern to the meteorologist is the ability of clouds to produce a large variety of precipitation forms. Depending on where you live, you are probably pelted from time to time with rain, snow, sleet, hail, snow pellets, or various combinations thereof. Changing conditions in the clouds and the air surrounding them control which of these forms of wetness will have its turn with us next. Because precipitation directly affects—sometimes with disastrous consequences—transportation, shelter, recreational activities, and agriculture, scientists are trying to learn as much as possible about the causes and possible control of precipitation from clouds. Much remains to be learned for complete understanding of these fascinating meteorological phenomena.

INVESTIGATION 1. (E)

Making a Cloud Model

MATERIALS NEEDED:

Large wide-mouth glass jar, 6–8 inches diameter
Rubber or cork stopper (3-hole) to fit
Styrofoam or foam-rubber insulation formed to fit outside of jar
Air thermometer, prepared by blowing a ¾-inch bulb on end of a soft glass tubing, 12 inches long
Mercury for bead, 1 gram
2 6-inch sections of rubber tubing, ¼-inch diameter
2 3-inch sections of glass tubing, ¼-inch diameter
12-inch glass tube bent in form of U to form manometer
Cardboard scale
Water
Flashlight

Our cloud or fog model will be contained within a glass jar. We can call this either a cloud model or a fog model, because fog and clouds are essentially the same. Both are composed of tiny water droplets (ice crystals if the temperature is low enough) in suspension in the air. The main difference between them is their location, clouds being formed at some distance above the ground and fog generally hovering close to the ground surface.

Consider for a moment the purpose of a model. It need not be a physical apparatus at all, but a mental model—an hypothesis or theory that uses some kind of imaginary construct, frequently supported by mathematics. In either case, the model is a device to help one understand.

The two models described in this chapter attempt to simulate on a very small scale a meteorological event, such as a cumulonimbus cloud. The model method has drawbacks. To study the dynamics of cloud formation, for example, we will induce convective currents in a tank of water. But, of course, water is much denser than air, and consequently the relationships between the forces at work and the movements of the material being studied are not an exact representation of what really happens in the atmosphere.

Nevertheless, models have the enormous advantage of being small and controllable. Experiments can be repeated any number of times, and data can be collected under a variety of conditions. Controlled experiments allow one to isolate experimental factors and apply them one at a time to note their specific effects. And most important, these models, even with their imperfections, will help you to understand the basic concepts that will facilitate your study of the atmosphere at large.

Materials needed for this experimental model include a large glass container, preferably a one- or two-gallon jar, with a wide mouth and a rubber or other tight-fitting stopper. You might be able to obtain a jar of these dimensions from a neighborhood restaurant or drive-in, or any other establishment that buys cleaning liquids or foodstuffs in large, economy-size containers. Scientific supply companies sell bottles of various sizes. If you cannot find a jar with a rubber stopper, make your own stopper out of a piece of styrofoam or "foam-board," obtainable in art supply or dime stores. This material is easily cut with a scissors or small knife and, if fitted tightly to the jar mouth, will be an adequate substitute.

An air thermometer with a bulb about ¾ inch in diameter will be inserted in the jar to measure small temperature changes. The working fluid in an air thermometer is air instead of alcohol or mercury. The air quickly responds to changes in temperature by expanding or contracting, causing a small bead of mercury to ride up or down in the thermometer's glass tube.

Construct an air thermometer by heating the end of a piece of ¼-inch diameter glass tubing in the flame of a propane torch or Bunsen burner until it closes. The tube should be long enough to extend about ¾ of the length of the jar you will be using, allowing an extra inch or two to extend through the stopper. Then carefully blow a bubble on the closed end of the tube by heating the glass to softness, removing from the flame, and blowing gently through the tube. A little practice here will result in a bubble about ¾ inch in diameter, which will make a suitable air thermometer bulb.

After the thermometer bulb is cool, using a medicine dropper, place a large drop of mercury in the open end of the tube. As the bulb cools, the mercury droplet will slide partway down the tube and come to rest at some equilibrium point, and will be ready to record temperature changes.

To calibrate the thermometer, compare it with a standard laboratory mercury thermometer, graduated in Fahrenheit degrees. Place both in a cool room. After a few minutes, note the temperature registered by the laboratory thermometer and the position of the mercury bead in the air thermometer. Mark the air thermometer at the bottom of the bead by lightly scratching the tube with a triangular file, and record separately the number of degrees the scratch represents. Next, place both thermometers in a warm room and repeat the process. Now, carefully put light scratches on the air thermometer tube to correspond to each degree of difference between your highest and lowest readings. Check your calibration with the standard thermometer in other situations.

A water manometer can be attached outside the jar with a rubber tubing connection to the interior. This device will measure small pressure differences. As the air pressure inside the jar decreases, atmospheric pressure will push the water in the manometer tube downward on the open side of the manometer and upward on the side that is connected to the interior of the jar. The difference in height between the two water levels is a direct measurement of pressure difference between the atmosphere and the interior of the jar.

To construct the water manometer, bend a length of ¼-inch diameter glass tubing into a U shape by gently heating it over a Bunsen burner. You will need to calibrate your manometer in small pressure units, such as millimeters of water. Attach a millimeter scale from a piece of graph paper to one arm of the manometer. As an approximation, 10 millimeters of water height is roughly equal to 1 millibar or 0.029 inch or mercury pressure. For comparison, daily pressure changes in the atmosphere are of the order of 25 millibars or 0.725 inch of mercury.

Assemble the several components of the cloud chamber, according to the diagram in Figure 1. The stopper must have 3 holes to fit snugly around 1) the air thermometer tube, 2) a

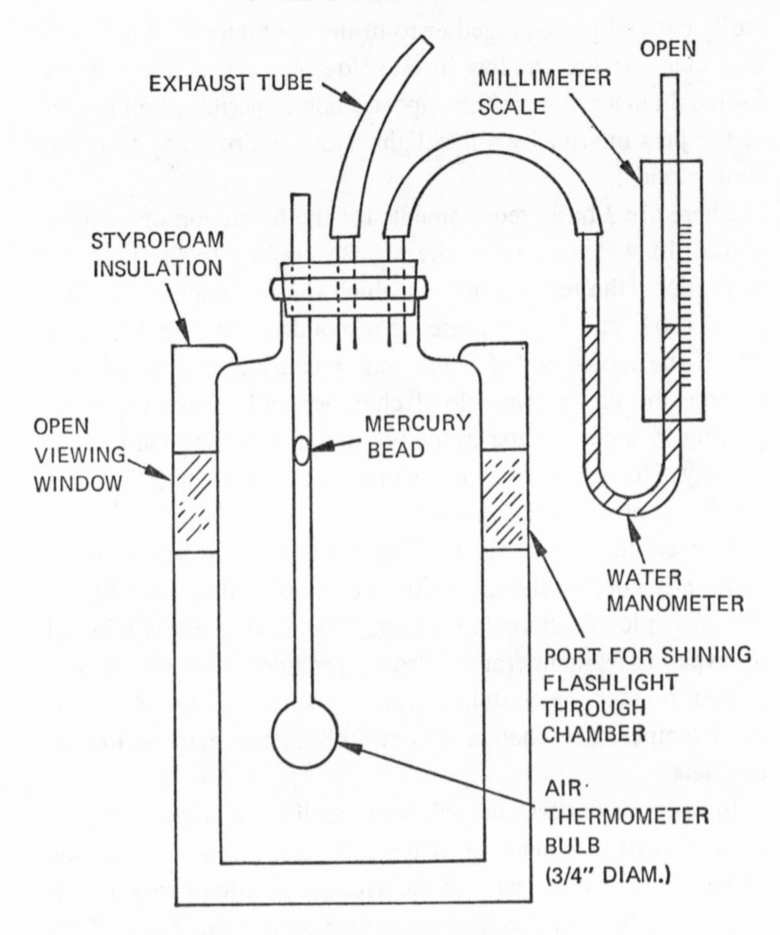

Fig. 1. Cloud Chamber

short length of glass tubing, to which you will attach about 12 inches of rubber tubing—this will serve as the exhaust outlet, and 3) a short length of glass tubing, which will be connected to the water manometer by another piece of rubber tubing.

For best results, the entire jar should be enclosed in some kind of insulating material, such as styrofoam, which can be easily cut and pasted together to fit the container. Cut a 3-inch diameter viewing window in one side of the insulation and a 2-inch diameter hole in the opposite side to permit illumination of the jar's interior by a flashlight. Now you're ready to make some clouds!

There are 3 basic requirements for the formation of a fog or cloud: 1) water vapor in the air, 2) cooling to the point of saturation (the temperature at which the air holds all the vapor it can), and 3) the presence of condensation nuclei, upon which the molecules of water vapor can begin to condense. Charge the air in your cloud chamber with water vapor by placing ½ inch of water in the bottom and shaking the jar vigorously. How does this insure that water vapor gets into the air? What process takes place?

To start the apparatus working, there must be some means of *rapidly* cooling the air inside the jar. For this, we will use the principle of adiabatic cooling. Whenever a gas is allowed to expand, its temperature drops, provided no heat is permitted to enter the container from the surroundings. An adiabatic temperature change occurs without the gain or loss of any heat.

In order to understand adiabatic cooling, it helps to have a clear idea of the meaning of temperature. Suppose we define temperature as a measure of the average kinetic energy, or energy of motion, of the molecules of a substance. Then, if we increase or decrease the average kinetic energy of the molecules, we raise or lower the temperature, respectively.

Think of a gas inside a cylinder containing a piston that can move back and forth, as shown in Figure 2. If the piston moves outward, the space inside the cylinder gets larger and the gas expands. Molecules that are moving toward the piston will collide with it and bounce back with *less* kinetic energy,

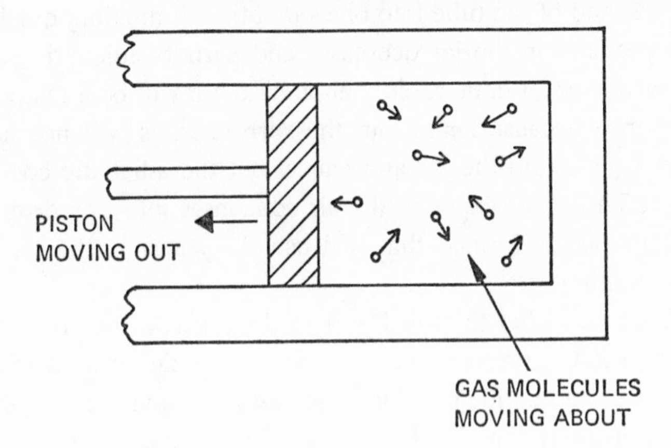

Fig. 2.

because the piston is moving in the same direction as they are. In the same way, a tennis ball thrown against the back of a truck that is moving away from you will bounce back slower than if the truck were standing still. As a result, the molecules of gas inside the cylinder have less average kinetic energy than before and, by definition, their temperature is lower.

How does this apply to the atmosphere? There are several ways by which air expansion can take place. Perhaps the most obvious is when horizontally moving air is forced to go up over a mountain range that stands in its way. As it rises, it enters a region where atmospheric pressure is less than near sea level. Thus, the rising air expands because it is not constrained as much by the surrounding atmosphere and, as it expands, it cools adiabatically, just as did the air inside the cylinder.

To get back to our cloud chamber—you should now be able to think of a way to cool the gas (air) inside the jar. This is where the exhaust tube comes into use. I have found that one can produce a good cloud or fog inside the jar by putting the free end of the tube into one's mouth and inhaling quickly. The pressure in the jar decreases and permits adiabatic cooling of the air. If done rapidly enough, there will be a temperature drop because heat from the surroundings will not have time to flow into the jar and cancel out the adiabatic cooling effect. The water vapor in the air condenses into tiny droplets and a cloud is formed that will usually persist in the jar for several minutes.

Remember that the third part of the formula for producing a cloud is the presence of condensation nuclei around which the water can condense. Air normally has some condensation nuclei in it. If the cloud you made by the method described above is not sufficiently dense, you might admit some additional condensation nuclei into the jar. Dust, soot, smoke, and salt particles are some of the condensation nuclei found in the atmosphere. For this experiment, the smoke from a match works well. Take a breath of air from the jar, pinch the exhaust tube, hold a lighted match near the opening, and release the tube. On your next trial, you will find a much denser fog produced than before the smoke particles were admitted.

Now that you have a working model of a fog or cloud, you will probably want to begin studying its action in greater detail. Use the following questions as guides to your investigation of the nature and formation of clouds.

1. Can you produce a cloud *without* introducing smoke into the chamber?
2. Can you use some other kind of condensation nuclei? Try substances like dust, flour, carbon black, lycopodium powder (a powder made from plant pollen, available in a chemistry laboratory). How do the results compare with the use of smoke?

3. Can you make a cloud if you fill the jar with filtered air, such as you might obtain by bubbling ordinary air through water? To obtain filtered air, temporarily remove the manometer tube and the exhaust tube from the stopper and replace each with a straight glass tube that extends to the bottom of the chamber. To the outside end of the new exhaust tube, attach a 2-foot length of rubber tubing and allow it to hang downward toward a sink or floor drain. Fill the chamber completely with water. Suck on the end of the exhaust tube until water flows out of the chamber by siphon action. Air will enter through the inlet tube where the manometer was formerly attached and bubble up through the water, effectively filtering the air. When all the water has been siphoned out, reattach the manometer tubing and perform your cloud-making experiments in the filtered air.
4. Can you devise a method to collect and examine the cloud droplets under a microscope? (See the experiment on collecting cloud droplets.) Before producing a cloud, mount a vaseline-coated microscope slide inside the jar. Remove and examine it carefully after each cycle of expansion and cooling.
5. How much does the temperature change inside the jar with each adiabatic cooling cycle? Your air thermometer will have to be calibrated carefully for this investigation. What is the minimum temperature change needed to produce a visible cloud?
6. How much does the air pressure change within the jar with each expansion cycle? Do you conclude that temperature change is more or less important than pressure change in producing a cloud? What influences your decision?

7. Why is an air thermometer used in this experiment rather than an ordinary liquid (mercury or alcohol) thermometer? Remember that air is very quick to react, and changes its volume greatly with small changes in temperature.
8. What would happen if you produced the necessary cooling by introduction of dry ice to the interior of the jar? Can you produce an ice-crystal cloud? If so, how does it differ from a water-droplet cloud?

INVESTIGATION 2. (E)

Making a Cloud Atlas

MATERIALS NEEDED:

Camera, any type
Small spiral notebook
Loose-leaf notebook—heavyweight paper for mounting photographs
Comparison photographs of clouds (e.g., International Cloud Atlas or Cloud Code Chart, U. S. Department of Commerce, Weather Service, U. S. Government Printing Office, Washington, D.C. 20231

One of the most pleasant ways to learn something about the workings of the atmosphere is to begin a thorough and systematic study of clouds through photography. Development of your own cloud atlas can be an interesting hobby, a source of repeated pleasure, and an activity that calls for no more equipment than an ordinary camera.

At first, you may feel you are faced with a seemingly impossible task because of the apparently endless varieties of cloud shapes and sizes. Gradually, however, your careful observations will reassure you that clouds do fall into categories that repeat themselves, and that the categories have limited types within them.

A cloud classification system was suggested by Luke Howard in 1803. The system, based upon structure and composition, is still used today, with the addition of altitude estimates. Three fundamental classes are stratus (layered), cumulus (heaped), and cirrus (fibrous). Combinations of these types are reflected by their names. Stratocumulus clouds, for example, are layered but have puffy or wavy surfaces resembling small cumulus. Cirrostratus clouds are wispy, fibrous clouds in layers at high altitudes.

The major categories in the classification system in use today are shown in this chart:

Cloud Type	*Description*	*Height Range*
	STRATIFORM OR LAYERED CLOUDS	
Cirrus	Layered, ice-crystal clouds. Temperatures below −25° C	High Bases above 20,000 ft.
Cirrocumulus		
Cirrostratus		
Altostratus	Fibrous or small cumuliform	Middle
Altocumulus	clouds. Temperatures from 0 to −25° C	Bases from 6,500 ft. to 20,000 ft.
Stratocumulus	Gray-colored, layered, wavy	Low Bases below 6,500 ft.
Stratus	Uniform, featureless	
Nimbostratus	Amorphous, dark gray rain clouds, 1,000 ft. to 2,000 ft. Temperatures more than −5° C	

Cloud Type	*Description*	*Height Range*
	CUMULIFORM OR HEAP CLOUDS	
Cumulus	Detached, puffy clouds of fine weather	Extend from 2,000 ft. to 20,000 ft.
Cumulonimbus	Heavy, dark clouds with extreme vertical development, generally giving showers or thunderstorms	Extend up to 40,000 ft. or higher
	SPECIAL TYPES OF CLOUDS	
Fracto clouds	Fragmented cumulus, stratus, or nimbus clouds	Low
Castellanus	Miniature, turreted, heap clouds signifying thundery weather	
Lenticular and wave clouds	Occur in mountainous areas in oscillatory airflow	Low and middle levels

You will probably very quickly be able to recognize the three main cloud types, but will have some trouble initially identifying those clouds that are combination varieties. Published collections of cloud photographs can be a great help as you gain experience. There are many fine books about clouds available; check your library. The *International Cloud Atlas* identifies the standard types, and you may wish to begin your own atlas by sighting and photographing the types the *Atlas* lists.

Some suggestions for pursuing this hobby have been described by Donald and Virginia Lokke in the August 1961 issue of *Weatherwise,* a magazine published by the American Meteorological Society. The equipment need not be expensive. A single-lens reflex camera, a small box camera, or even the cartridge-loading Instamatic[R] can be used to produce excellent cloud photographs. The Lokkes successfully photographed clouds with three different types of camera: An Agfa box camera, a Kodak Vigilant, and a Leica Model A. Verichrome and Verichrome Pan films were used in the box camera and in the Vigilant, and Plus-X in the Leica.

Naturally, a camera that permits regulation of shutter opening and speed will allow a wider range of experimentation in picture-taking. If you are using such a camera, you can gradually refine your technique by recording with each photograph the frame, or picture, number and the shutter opening and speed used, and comparing them later with the printed results. A yellow filter—sometimes called a "sky filter"—is often used with black-and-white film to eliminate haze and achieve greater definition of the subject being photographed. Inexpensive filters can be purchased from camera shops and laboratory supply companies, and an excellent pamphlet on how to use these helpful devices is available at small cost. A 35-millimeter slide camera will produce colored transparent slides for projection, particularly valuable for classroom use or for special slide showings for hobby clubs and the like.

For development of your pictures, you may rely upon your local camera shop or drugstore developing service. File the properly labeled negatives in dustproof envelopes. Later you may wish to try your hand at developing your own prints, making enlargements, etc. Selected negatives can then become the basis for a more refined collection of cloud photographs for your growing atlas.

Keep a complete record of information concerning photographic details and cloud type. Use a small spiral notebook or a set of 3-inch-by-5-inch cards as you are photographing, and later record the data alongside the mounted cloud photograph. A loose-leaf notebook with sheets of heavyweight paper is a good atlas base and will permit additions and deletions as you acquire more photographic examples of the clouds you wish to illustrate.

You will probably want to note:

Cloud photograph number
Date
Time
Location
Camera direction
Film
Camera—shutter opening and speed
Comments—filter, etc.
Field identification of cloud
Weather and landscape conditions
Estimated cloud heights and speed (see Investigations 4, 5, and 6)

The value of these kinds of data will become apparent with further experience. You will want to improve your technique and check the results of varied camera settings. As a source of scientific information for later study and analysis, your complete records will form an invaluable resource. The atlas can be embellished with pertinent newspaper clippings, such as weather maps, jotted notes from the day's weather report, or whatever other information you think will help to round out the meteorological picture.

Besides a camera, the main requirement for this fascinating study is a constant readiness to snap pictures when the opportunities present themselves. Clouds are shifting masses of often spectacular appearance. Quite often, striking examples of cloud formations last only a brief time and, without a camera at hand, the potential photograph is gone forever!

INVESTIGATION 3. (E)

Model of a Cumulonimbus Cloud

MATERIALS NEEDED:

Plastic aquarium, transparent, about 5–10 gallon capacity
Large metal tray, approximately 4 inches wider and longer than base of aquarium, with 2-inch sides
Small alcohol burner
One gallon of diluted milk (1 cup to 15 cups water)
Ice cubes
Bricks or wood blocks for support
Thermometer
Pycnometer

To the meteorologist, few natural phenomena surpass the beauty of a growing cumulus cloud on a sunny afternoon, as it progresses from a fluffy white mass on the horizon to the menacing, noisy, showering cumulonimbus cloud it may ultimately become. What forces are at work to cause this exciting display? Why is the process repeated so frequently, and with almost identical results each time? Is the cumulonimbus cloud a cause or an effect in the battle of the atmospheric elements? There's a simple way to demonstrate the formation of a cumulonimbus cloud, but first a brief review of the principle of convection will be helpful.

Cumulus, altocumulus, and cumulonimbus are all convective clouds, caused by differences in density in the atmosphere and the resulting vertical motions that occur. The basic force behind these motions is gravity. Air that is heated from below, either by being in contact with a warm ground surface or body of water, or by the influx of warm, moist air from surrounding areas, becomes buoyant and begins to rise, pushed upward by cooler, denser air in the vicinity, just as a hot balloon is caused to rise.

As the lighter air ascends, it expands and cools adiabatically (see Investigation 1 for a discussion of adiabatic cooling) until its dew point, or the temperature at which the air is saturated, is reached. At this altitude condensation begins, and the visible cloud makes its first appearance. The cloud may continue to grow if the supply of water vapor in the rising air is adequate, and, as it grows, there is a tendency for new, drier air to be drawn in from the sides. This is called "entrainment," and is one reason why strong surface winds may develop in the vicinity of large cumulonimbus clouds. The cloud seems to act like a huge vacuum cleaner, rapidly drawing in large quantities of air (see Figure 3). Since this new air is drier and cooler than that in the center of the cloud, with its rapidly rising column of warm air, it has an inhibiting effect on the growth of the cloud. Thus, in order for a cumulus cloud to grow and expand into a large cumulonimbus cloud and eventually become a thunderstorm, there must be a large enough supply of warm, moist air available to counteract the drying effect of the air that is entrained from the sides. A thin, narrow cell of warm, moist air would be so mixed with dry air that it would lose its buoyancy. On the other hand, a broad-

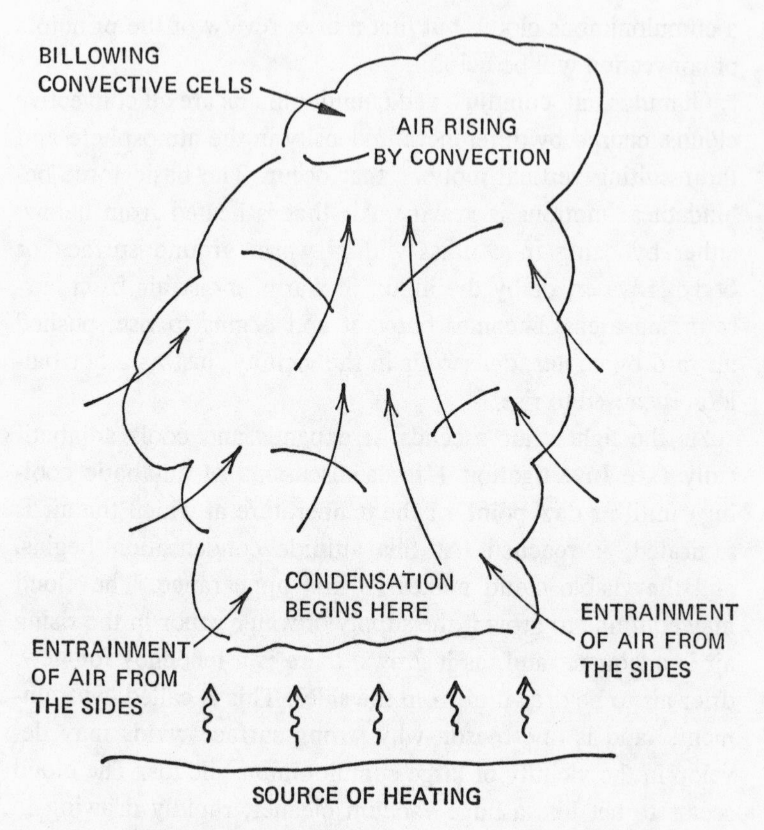

Fig. 3. Actions in a Convective Cloud (cumulus stage)

based convective cloud might have sufficient diameter that the entrained air would affect only the edges, leaving a core of warm air at its center still capable of rising buoyantly. If you observe cumulonimbus clouds, you will notice that those that grow highest are also large in diameter, with broad bases. Consider this point when you compare the results obtained from your model with what you observe in nature, and remember that your model uses a small, local source of heat.

For this model, you will need a large, preferably ten-gallon plastic aquarium and a tray of ice cubes big enough to set the aquarium in. You will also need a gallon of much diluted cold milk (about 1 cup of milk to 15 cups of water), and a Bunsen burner or propane torch. As you can guess, we will induce a "cumulus cloud" of milky water to develop in a tank of clear water by raising the temperature of a portion of the "ground" surface.

Fill the aquarium about four-fifths full of cold water and set it in the large tray of ice cubes. Let the water come to rest. Carefully pour in the much-diluted cold milk through a thistle tube or long-necked funnel to the bottom of the tank, to make a layer about 1 inch deep (see Figure 4). With the

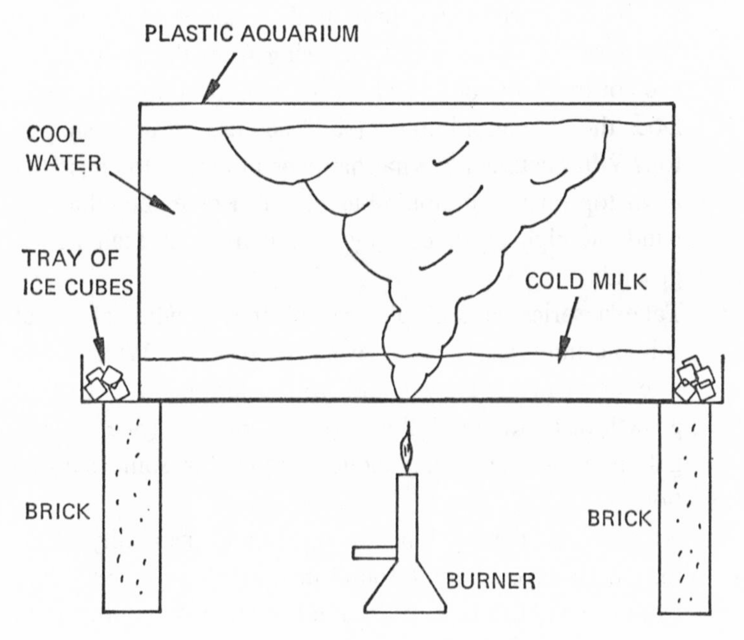

Fig. 4. Model of a Cumulonimbus Cloud

burner, apply heat gently to the milk in the bottom of the tank and observe the growth of the "cumulonimbus cloud."

You may wish to arrange a lighting system so that the image of the growing cloud can be projected on a screen for classroom demonstration. Place a slide projector or carbon arc light for a light source about 4 feet from the aquarium. Shine the light through the aquarium so that a shadow image of the "cloud" will be projected on a wall or screen.

1. Does the cloud go through stages of development? Note that there is an important difference between your model and the development of a cumulonimbus cloud in nature. The model involves a "point" heat source applied only at the bottom, while a cumulonimbus has a collection of "bubbles" of warm, moist air rising from the cloud base, each of them giving up its heat by condensation as it rises.
2. Does the "cumulonimbus cloud" acquire an anvil-shaped top? What causes it? What happens in nature to cause the anvil top on a cumulonimbus cloud? Remember that the winds at high altitudes may be much faster than at the ground.
3. Take a series of still pictures of the growing cloud at suitable intervals, such as every 10 seconds. What is the rate of growth? How could you determine the rate of growth of a cumulonimbus cloud in nature? (See Investigation 4 on measuring cloud heights for some suggestions.)
4. An excellent description of stability and instability of atmospheric layers can be found in *Watching for the Wind*, by James G. Edinger. What kind of stability is present in your model? What kind of atmospheric stability is usually present when a cumulus or cumulonimbus cloud

forms? Air is considered *stable* when it has little tendency to rise vertically or to overturn. When air near the ground is warmed, it becomes less dense than the surrounding air, tends to be forced upward, and the air is said to be *unstable*. Unstable air, as you know, tends to encourage development of cumulus and cumulonimbus clouds.

5. What is the temperature lapse rate (change of temperature with height) in the aquarium?
6. What is the density of the original cold water and of the dilute milk solution, before and after heating, in the aquarium? To determine density, you will have to borrow two measuring devices from a physics or chemistry laboratory—a pycnometer and a balance. Fill the pycnometer, a small container of known volume, with clear water from the aquarium and weigh it with the balance, which will tell you the mass of the contained water. From the values of mass and volume, the density can be computed: Density is equal to mass divided by volume. By using a pipette (an open glass tube that you can fill with liquid by inserting the tube in the fluid and stopping the free end with your finger), or, carefully, by using a medicine dropper, remove a sample of the milk solution from the bottom layer in the aquarium. Apply the same method of measurement to the dilute milk solution, before and after it is heated. What do these density measurements indicate about pressure differences in the aquarium? And what connection do they have to the growth of the "cloud"?

What techniques of study can you use to investigate other phenomena with your model of the cumulonimbus cloud? Can you simulate other types of clouds with this apparatus?

What type of cloud is represented after the cumulonimbus dissipates or spreads out? Remember that any cloud that is in layers or "strata"—such as stratocumulus, altostratus, or cirrostratus—shows a stable condition of the atmosphere.

INVESTIGATION 4. (M)

Measuring Cloud Height

MATERIALS NEEDED:

Materials for cloud height measurer:

Plywood square 12 inches by 12 inches by ¾ inch
3 screws, 3 inches long
Protractor
Magnetic compass
Plywood board 12 inches by 6 inches by ¾ inch
Wooden stick, 12 inches by 1 inch by ½ inch
Metal eyelet screw, 1 inch long
Pin, ⅛-inch head

Materials for sling psychrometer:

2 Fahrenheit thermometers (−10° F to +100° F range)
Plywood block, 10 inches by 4 inches by ½ inch
6 small screws, ½ inch long
Coat hanger wire, 16 inches long
Wooden dowel, 1 inch diameter, 6 inches long
2 small washers
1 screw, 2 inches long
Small square of cheesecloth, 2 inches by 2 inches

Have you ever watched cumulus clouds on a fine summer day and wondered how high they were? Perhaps you saw small planes appearing and disappearing as they flew through them, and from this concluded that the clouds were probably several thousand feet above the ground.

You'll learn here two different ways to measure cloud heights from the ground. The first method employs some basic trigonometric calculations on a theoretical triangle formed by you, at one sighting point, an associate at sighting point two, and a cloud. The second method makes use of a simple device to measure atmospheric moisture. Both systems are most successful when used to measure the heights of cumulus clouds.

According to the system of classification used by the *International Cloud Atlas,* clouds fall into four major altitude categories: Low (bases less than 6,500 feet above the ground), Middle (bases 6,500 feet to 20,000 feet above the ground), High (bases more than 20,000 feet above the ground), and clouds of Vertical Development (extending from near the ground to above 20,000 feet). Because of the difficulty of measuring angles accurately with the instruments described in the first method, your results will be best when finding the heights of clouds in the Low category, which includes cumulus and stratocumulus. Stratus clouds are more difficult to measure because of the indistinct character of the base of such clouds and the problem of locating a suitable point on which to sight.

To measure cloud heights using the method I am going to describe, you will need to enlist the aid of a helper. You will also need to construct two identical instruments for measuring angles of elevation. A diagram of the construction details is shown in Figure 5.

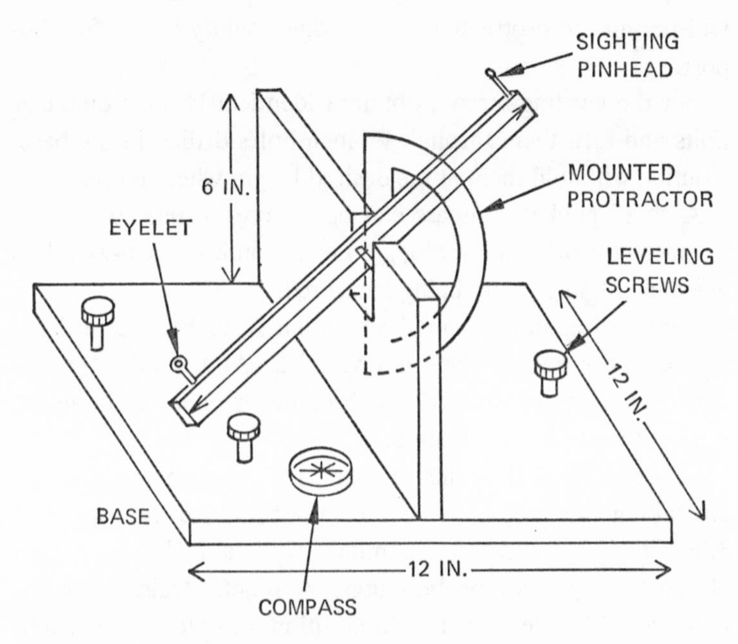

Fig. 5. Cloud Height Measurer

With a board 1 inch thick and twelve inches square as a base, secure a second board 1 inch by 6 inches by 12 inches to form the upright. Attach the upright board to the base with 1-inch screws from the bottom. Saw a notch 1 inch wide and 2 inches deep in the upright, as shown, and mount a narrow stick (1 inch by ½ inch by 12 inches) on a pivot inside the notch. The pivot can be a small nail hammered through the middle of the stick (6 inches from either end) and fitted into small holes drilled into the side of the notch.

Mount a small metal or plastic protractor to one side of the notch, with its center line at the same height as the pivot. A thin groove sawed vertically along one side of the upright will permit the protractor to be wedged tightly into it for support.

For the leveling screws, obtain 3 identical 1/4-inch diameter bolts and turn them through 1/4-inch holes drilled in the base. Friction will hold them at the desired height when in use.

A small pocket compass can be set on the base of the instrument for orientation along a north–south or east–west line when cloud heights are being measured.

For sighting, insert a small pin with a 1/8-inch diameter head in the upper end of the sighting stick. At the other end, a small screw eyelet with a 1/4-inch diameter opening should be attached as shown in Figure 5.

Construct 2 of these instruments, and you are ready to begin taking measurements. Decide upon a base line of about 3,000 feet (or close to 6/10 mile), preferably level and unobstructed by trees or buildings. A quiet, straight country road would be ideal. If it is, in addition, oriented north–south or east–west, it will be easier to calculate angles should you decide to measure cloud directions as well as heights. The base line can be measured accurately enough by use of the odometer in a car. Check the distance by driving the car over the base line twice and averaging the results.

Or, without a car, use the speedometer on a bicycle or walk off the distance and estimate the length of the base line with an inexpensive pedometer. You may be able to set up your sighting instruments along a base line that is premeasured—for example, property boundaries for which you can ascertain the length without actual measuring. Whatever method you use, come up with as accurate an estimation of the distance as you can.

If you don't live near flat, open country, or a reasonably deserted stretch of beach, you may have some trouble finding an unobstructed area in which to work. Your measurements will still work if the base line is reduced up to one half, or 1,500 feet (about 6 city blocks).

Pick a day when there are plenty of clouds from which to choose. It won't matter too much if the clouds are moving rapidly because you and your assistant will take simultaneous readings on some common point of a particular cloud moving overhead. The most convenient way to synchronize your readings is to use a walkie-talkie. If you are in the city, you might use suitably located phone booths. When you are each in position at opposite ends of the base line, you will be able to see the same cloud and confer about what point both of you will sight on.

If you do not use a walkie-talkie, it will be necessary to synchronize your watches and decide upon a specific time at which you will take your readings. If a level road is the base line, one observer can be dropped off at one end and the other can go to the point 3,000 feet away, having previously decided to measure a particular cloud at, say, the tip nearest observer one, and then both observers can immediately take their readings at the predetermined time. Or, should you be within sight of each other at opposite ends of the base line, send a prearranged signal—a large, red cloth waved at one end, for example—when the readings are to be taken.

Set the measuring instrument on an old box or other raised object, sight through the screw hole on a line with the pinhead, and note the degrees indicated on the protractor. Try to take 4 or 5 trials on the same type of cloud, and average the final results.

Following is a sample measurement made one fall day on some clouds over Denver. The diagram (Figure 6) shows the

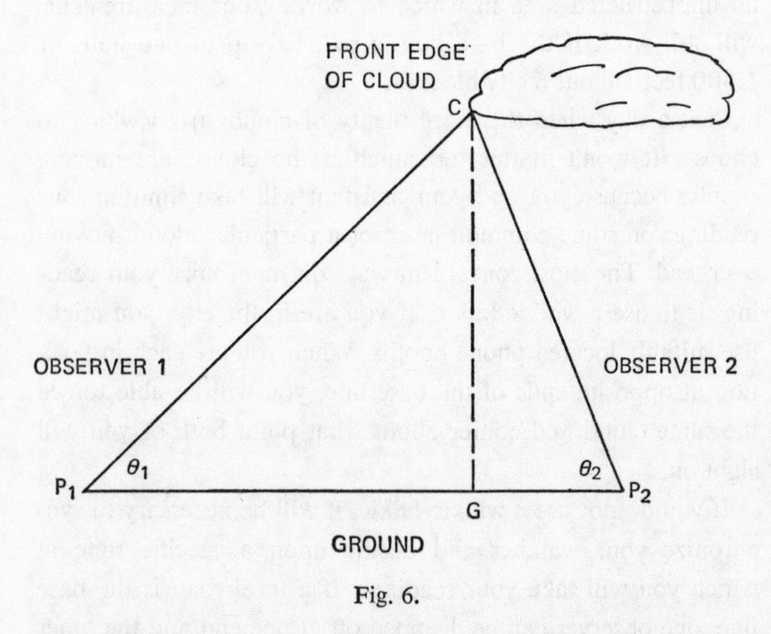

Fig. 6.

required angles and base-line measurements needed for calculation of the cloud height. The problem was solved as follows:

$P_1P_2 = 3{,}000$ feet
$\theta_1 = 69°$; $\theta_2 = 82°$
* Tangent $\theta_1 = 2.605$; Tangent $\theta_2 = 7.115$

(1) $\text{Tan } \theta_1 = \frac{CG}{P_1G}$; $\text{Tan } \theta_2 = \frac{CG}{P_2G}$

(2) $CG = P_1G \text{ Tan } \theta_1$; $CG = P_2G \text{ Tan } \theta_2$

(3) $P_1G \text{ Tan } \theta_1 = P_2G \text{ Tan } \theta_2$

(4) $P_1G \text{ Tan } \theta_1 = (3{,}000 - P_1G) \text{ Tan } \theta_2$

(5) $P_1G \text{ Tan } \theta_1 = 3{,}000 \text{ Tan } \theta_2 - P_1G \text{ Tan } \theta_2$

(6) $P_1G (\text{Tan } \theta_1 + \text{Tan } \theta_2) = 3{,}000 \text{ Tan } \theta_2$

(7) $P_1G = \frac{3{,}000 \text{ Tan } \theta_2}{\text{Tan } \theta_1 + \text{Tan } \theta_2}$

(8) $P_1G = \frac{(3{,}000)\ (7.115)}{2.605 + 7.115}$

(9) $P_1G = \frac{21{,}345}{9.720} = 2{,}196$ feet

(10) $CG = P_1G \text{ Tan } \theta_1 = (2{,}196)\ (2.605)$

(11) $CG = 5{,}720$ feet (height of cloud)

NOTE: The tangent of an angle is the ratio of the side opposite the angle to the side adjacent to the angle, in a right triangle. Refer to Appendix B for tangent tables.

The preceding measurements were taken on a cloud that passed directly over each of the observers. To measure the heights of clouds that do not pass directly overhead, the mathematics is a little more complicated, but requires nothing beyond basic high school trigonometry.

In this case you will again use a base line of 3,000 feet, but it will be necessary to measure not only the elevation angles θ_1 and θ_2 to the cloud from each end of the base line but also the azimuth angles δ_1 and δ_2. Elevation angles are measured from the horizontal to the line of sight of the cloud. Azimuth angles are measured from the north–south line (or east–west line if the cloud is east or west of you) to a point on the ground directly under the cloud (see Figure 7).

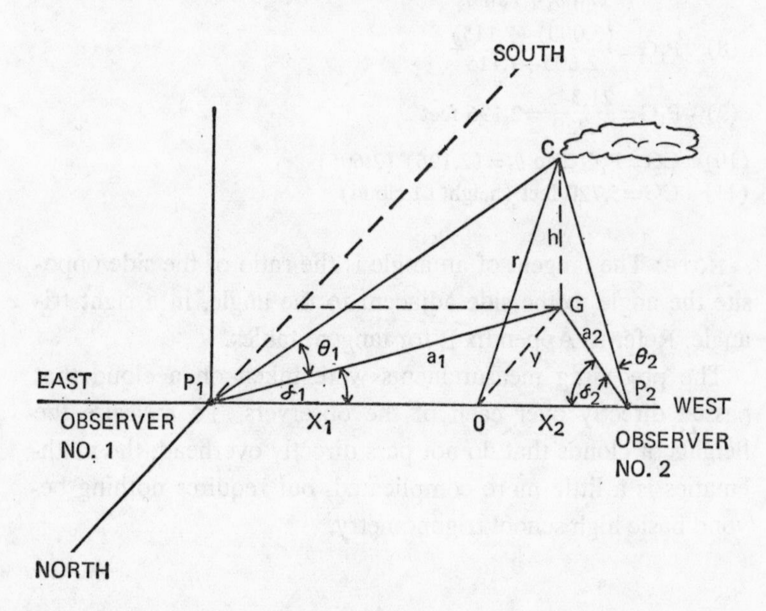

Fig. 7.

To find h, the cloud height:

(1) $y = a_1 \sin \delta_1 = a_2 \sin \delta_2$

(2) $a_1 = \frac{y}{\sin \delta_1}$; $a_2 = \frac{y}{\sin \delta_2}$

(3) $y = x_1 \tan \delta_1 = x_2 \tan \delta_2 = (3{,}000 - x_1) \tan \delta_2$

(4) $h = a_1 \tan \theta_1 = a_2 \tan \theta_2$

(5) Substituting,

$$h = \frac{x_1 \tan \delta_1 \tan \theta_1}{\sin \delta_1} = \frac{(3{,}000 - x_1)\ (\tan \delta_2)\ (\tan \theta_2)}{\sin \delta_2}$$

(6) Solving for x_1,

$$\frac{3{,}000 - x_1}{x_1} = \frac{(\sin \delta_2)\ (\tan \delta_1)\ (\tan \theta_1)}{(\sin \delta_1)\ (\tan \delta_2)\ (\tan \theta_2)}$$

(7) $$x_1 = \frac{3{,}000}{\frac{(\sin \delta_2)\ (\tan \delta_1)\ (\tan \theta_1)}{(\sin \delta_1)\ (\tan \delta_2)\ (\tan \theta_2)} + 1}$$

(8) $$h = \left[\frac{3{,}000}{\frac{(\sin \delta_2)\ (\tan \delta_1)\ (\tan \theta_1)}{(\sin \delta_1)\ (\tan \delta_2)\ (\tan \theta_2)} + 1} \right] \left[\frac{(\tan \delta_1)\ (\tan \theta_1)}{\sin \delta_1)} \right]$$

To illustrate the calculation by this method of the height of a cloud that does not pass directly overhead, let:

$$\delta_1 = 30°,\ \delta_2 = 40°$$
$$\theta_1 = 45°,\ \theta_2 = 50°$$

$$h = \left[\frac{3{,}000 \text{ feet}}{\frac{(.643)\ (.577)\ (1)}{(.5)\ (.839)\ (1.192)} + 1} \right] \left[\frac{(.577)\ (1)}{(0.5)} \right]$$

$$= \frac{3{,}000 \text{ feet}}{1.0742}\ (1.154)$$

$$h = 3{,}223 \text{ feet}$$

Become familiar with your instruments, take a number of sightings on different kinds of cloudy days, and consider some interesting questions. How do cloud heights change with time? Do all clouds of a given type have the same base height? Do you find that cloud bases or cloud tops are more nearly the same height?

Since cumulus clouds are caused largely by convection, during which a column of moist air is rising and cooling, there is a method for finding the heights of their bases that employs dew points, or the temperature at which condensation of water vapor would begin if the air were cooled to that temperature. For the dew-point method, you will require a sling psychrometer and a set of psychrometric tables for obtaining dew points. The Government Printing Office will supply a complete set of these tables (see the appendix for the correct address); for convenience, however, an abbreviated set of psychrometric tables is included in Appendix B.

A sling psychrometer can be constructed easily from 2 identical, inexpensive outdoor thermometers, available in any hardware store. Carefully cut away the lower ends of the holders so that the bulbs and about ¼ inch of the capillaries above them are free. Mount the 2 thermometers side by side on a thin board (½ inch by 4 inches by 10 inches) as shown in Figure 8. Wrap one bulb with a small piece of cheesecloth to

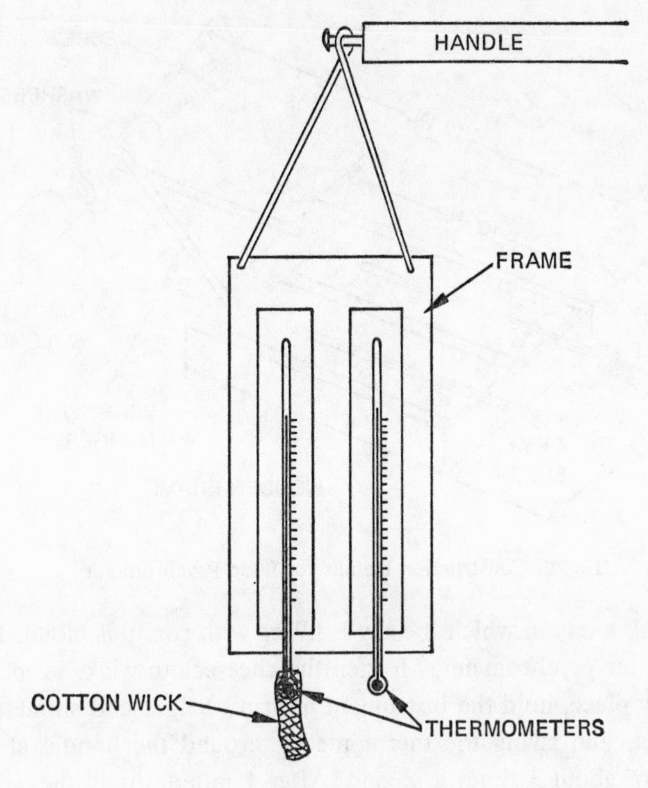

Fig. 8. Sling Psychrometer

form a wick, and suspend the board by a coat hanger wire from a short handle (construction details are shown in Figure 9).

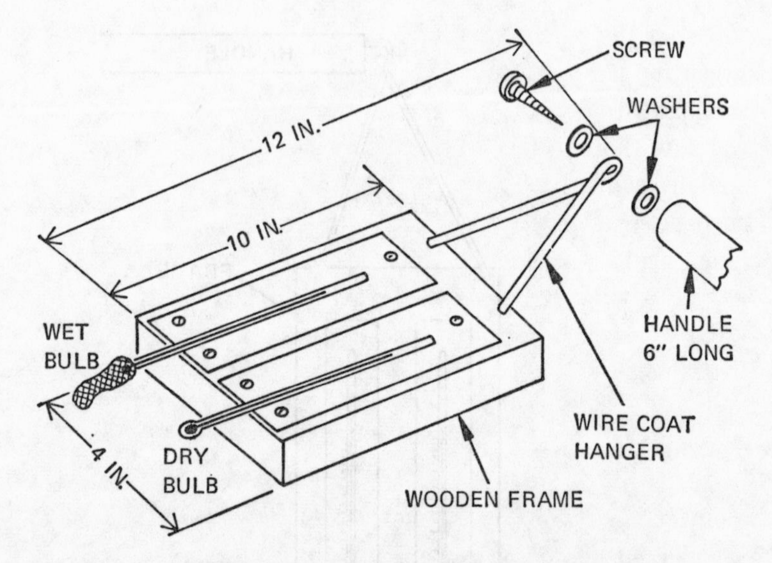

Fig. 9. Construction Details for Sling Psychrometer

Pick a day in which the sky is filling with cumulus clouds to use your psychrometer. Moisten the cheesecloth wick, select a shady place, hold the instrument at arm's length and shoulder height, and swing the thermometers around the handle at a rate of about 3 times a second. After 1 minute, read the wet-bulb thermometer quickly and then the dry-bulb thermometer, each to the nearest ½ degree. Continue to swing the thermometers until 2 successive readings show no temperature changes, and use these readings to find the dew point.

To use the psychrometric tables, compute the difference between the temperatures of the dry bulb and wet bulb on the sling psychrometer (t–t′). On the dew-point tables, read downward on the left side to the dry-bulb temperature and horizontally across the top to the number representing the difference between the dry- and wet-bulb readings. The number found at that intersection is the dew-point temperature in degrees Fahrenheit.

To what we already know about the formation of convective clouds, such as cumulus and cumulonimbus, add an additional piece of information. The normal rate at which the dew point decreases with height above the ground is different from the dry adiabatic lapse rate, the rate at which air normally cools, without the gain or loss of any heat, as it rises and expands in the atmosphere. Consider what happens as a parcel of air is caused to rise by convection on a warm summer afternoon. As long as condensation does not occur, the temperature of the parcel will decrease by adiabatic cooling at the rate of 5.5° F per 1,000 feet of rise. At the same time, the rate at which the dew point decreases for the same rising parcel of air is 1.1° F per 1,000 feet. The temperature of the air, therefore, will decrease faster than its dew point, and after a time the two will be the same. This is illustrated in Figure 10.

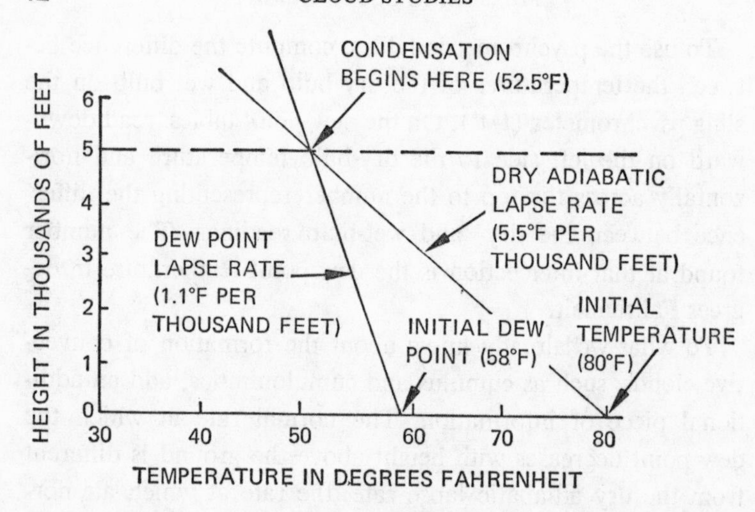

Fig. 10.

Suppose a parcel of air at the ground with a temperature of 80° F and a dew point of 58° F starts to rise in a convective cell. The difference between these two temperatures is 22 degrees. The rate at which the temperature overtakes the dew point is 4.4° F per 1,000 feet (5.5° F−1.1° F)/1,000 feet. Thus at a height of 5,000 feet (22° F÷4.4° F/1,000 feet= 5,000 feet), the two temperatures will be the same (52.5° F) and condensation will occur. This will be the base of the cumulus cloud that begins to form. Clouds will form (condensation begins) whenever the air reaches saturation or when the air reaches its dew point.

It can be seen that in this procedure we have a method for determining the height of the base of cumulus clouds. An equation that expresses the relationship used is:

$H = 227 (T_o - D_o)$, where

H is the height sought (in feet)

T_o is the actual Fahrenheit temperature of the air at the ground, and

D_o is the dew-point temperature of the air at the ground.

(For the derivation of this equation, see Appendix A.)

Other questions that you might investigate with your methods of measuring cloud heights are:

1. How do various points on the base of a single cloud vary in height?
2. Does a cumulonimbus cloud grow downward as well as upward? Remember that these clouds grow because of convection and difference in air densities.
3. Can you measure the rate of ascent of a cumulonimbus cloud top? How do different cumulonimbus clouds vary in ascent rates?
4. Find the vertical thickness of a cloud by this method. Why do some cumulus or cumulonimbus clouds appear darker than others? Which permit more sunlight to come through—thin ones or thick ones?
5. From your observations, can you tell whether a growing cloud is a small cloud expanding by original water droplets moving upward and outward, or growing by new cloud material forming at the surface of the cloud, as in crystal growth? Would new cloud material form around the edges of the main cloud if conditions were right for condensation of water vapor?

INVESTIGATION 5. (E)

Measuring Cloud Direction and Speed

MATERIALS NEEDED:

Small table or stool, 18–24 inches high
Magnetic compass
Circular mirror, 6 inches in diameter
Plywood, 9 inches by 9 inches by ¾ inch
3 screws, 2 inches long
Wooden dowel, ¼ inch by 3 inches
Paper circle, 9 inches in diameter

A weather observer is interested in cloud direction and speed because of the information it gives him about winds aloft. And frequently this information is rather surprising. For example, would you expect winds a thousand feet above the ground to be exactly opposite in direction and at twice the speed of those at the ground? In fact, the average velocity of the wind does increase with height above the ground. This effect is most noticeable in the first 100 feet, the velocity generally doubling between a height of 1½ feet and 33 feet, and increasing an additional 20 per cent to a height of 100 feet. Such a large variation is mainly due to the reduction in frictional "drag" with increased height above the ground. Along with the increased velocity, there is usually a decrease in turbulence also, although some surface eddies can affect the air several thousands of feet above the ground, particularly in mountainous areas.

On a day when noticeable winds are present at the ground, pay particular attention to their speed and direction. Then observe the clouds drifting above you. Are they traveling in the same direction as the winds at the ground? How can you obtain a reasonably accurate measurement of their direction and speed?

An instrument used for such measurements is called a nephoscope. The simple version you can construct following the directions in this section can be used very nicely in conjunction with the cloud height measuring techniques described in Investigation 4. As you will see, the formula for calculating speed includes cloud height. If you have become proficient in the use of the height measurer or sling psychrometer, add the nephoscope to your bag of equipment to make a complete analysis.

Obtain a circular mirror about 6 inches in diameter. Mount it on a flat piece of plywood about 9 inches in diameter, or 9 inches square, if you don't have a circular piece of wood. Glue a paper circle around the mirror and flat on the plywood surface. The circle should have a diameter of 7½ inches and should be carefully marked off into 360 degrees. In other words, the calibrated paper border should be about 1½ inches wide.

Glue or screw a sighting point, made of a ¼-inch dowel 3 inches long, to the plywood at the mirror's edge, as shown in Figure 11. Mark with a felt-tip pen, lipstick, or whatever, a small dot in the center of the mirror. Finally, screw 3 ¼-

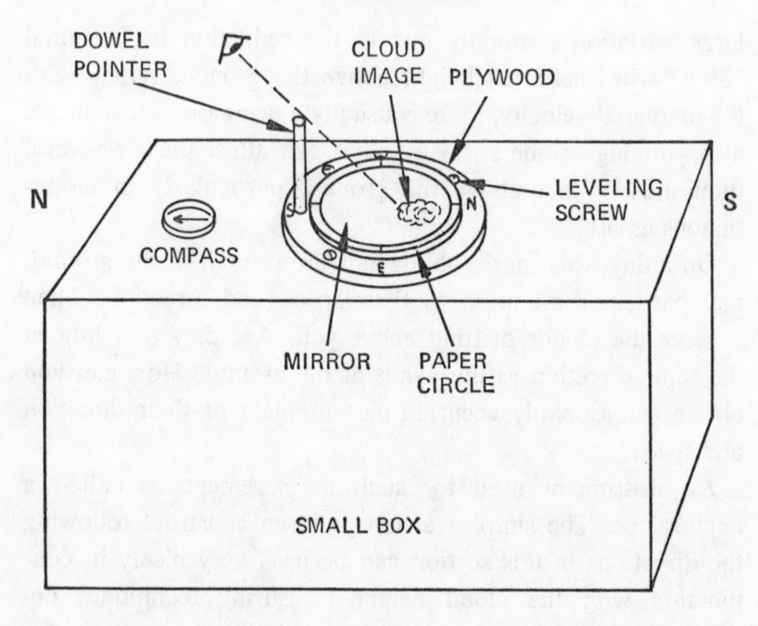

Fig. 11.

inch screws at 3 points around the plywood border. These should be long enough to extend through the base, and will be used as leveling devices when you are ready to work.

Set up the nephoscope outside on a box on reasonably level ground, and, using a small pocket compass, orient the instrument so that south on the paper circle is pointing north (see the diagram). Bring the nephoscope to approximate level by the 3 adjusting screws.

To determine cloud direction, locate a cloud image at the dot in the center of the mirror. Continue watching the cloud image by sighting over the sighting point, and note the angle at which the image disappears from the edge of the mirror. This represents the wind direction at the height of the cloud observed.

To determine speed, you'll need a stop watch, or a friend with a watch with a second hand who will do the timing as you watch the image. Without either, you can always count off seconds. Time the movement of the image from the center dot to disappearance at the mirror's edge. The formula for the calculation of cloud speed is:

$$\text{Speed} = \frac{\text{cloud height (feet)} \times \text{mirror radius (inches)}}{\text{height of dowel (inches)} \times \text{time (seconds)}}$$

NOTE: If the mirror radius and dowel height are the same, the equation reduces to:

$$\text{Speed} = \frac{\text{cloud height (feet)}}{\text{time (seconds)}}$$

Another way of applying the same technique to determine the speed and direction of winds at cloud level can be called the "direct vision nephoscope." In this case, you—the observer—move to keep the cloud sighted relative to a fixed point. Broken clouds at night may be sighted against the stars or moon; in the daytime, use a flagpole, light pole, or a prominent part of a building. To determine cloud direction alone, simply keep the eye stationary and watch the cloud with reference to the flagpole, or whatever you are using as your fixed point. To measure cloud velocity, it will be necessary to ascertain the time the cloud took to move through a certain angle with reference to the fixed point. Keeping the cloud sighted over the flagpole, move your position so as to keep a point on the cloud in line with the top of the pole. Drop a marker at the beginning of the observation and another at the end of a known time interval. Measure the distance you have moved during the observation and note the exact direction of movement. You will be moving in a direction opposite to that of the cloud, and you will be forming 2 similar triangles of which the angle points are eye level (at positions beginning and end) and pole, and cloud (at positions beginning and end) and pole (see Figure 12). Therefore,

$$\frac{D}{H} = \frac{d}{h}$$

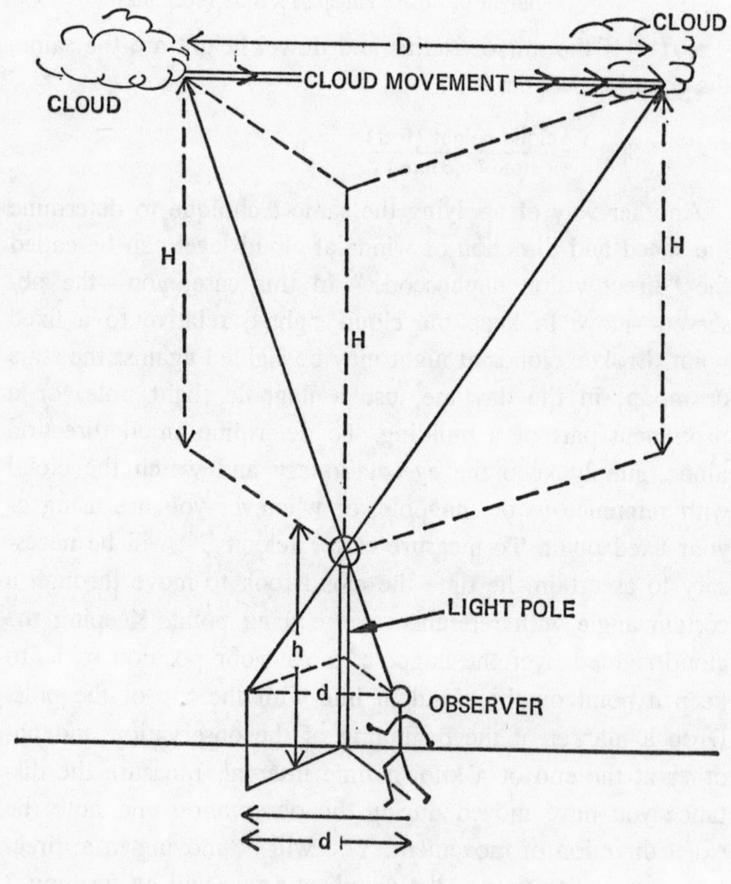

Fig. 12. Measurements Needed in Direct Vision Nephoscope

The speed of movement of the cloud is: $V=D/t$. Substituting this in the proportion, we get:

$$V=\frac{dH}{ht,}$$

where d is the distance moved by the observer in feet,
h is the height in feet of the pole (measured or estimated) from above the observer's eye level,
t is the time in seconds, and
H is the height of the clouds being measured from the top of the light pole. Because the height of the light pole is extremely small in comparison with the height of the clouds being measured, it is satisfactory to call H the height of the clouds above the ground.

Using your estimation of cloud height from the measuring techniques described in the previous section, feed your data into the formula and come up with how fast the winds at cloud level are traveling.

Some questions you might try to answer as you take measurements of speed and direction on moving clouds are:

1. Which types of clouds seem to move faster, high clouds or low clouds?
2. Can you use the nephoscope methods to obtain the rate of growth of a cumulonimbus cloud? Try it.
3. Determine the cloud direction and speed for 2 levels of clouds on the same day, and compute the wind shear between the 2 levels. Wind shear is the variation of wind speed and direction over a given vertical distance. If, for example, the lower layer of clouds had a velocity of 20 feet per second from the west, and the upper layer of

clouds (say, 5,000 feet higher) had a velocity of 60 feet per second from the northwest, the wind shear could be described as 60 fps—20 fps=40 feet per second per 5,000 feet, or 8 feet per second per 1,000 feet, clockwise, between the lower and the upper layers of clouds. In this case, wind shear is due to the shift in wind direction from west to northwest (clockwise) with the vertical ascent from the lower level to the upper level.

4. Observe some lenticularis clouds, such as stratocumulus lenticularis, or altocumulus lenticularis, and compute their height, speed, and direction of movement. How does the speed differ from ordinary stratocumulus or altocumulus clouds present on the same day? The lenticularis cloud is caused when air is set into vertical oscillatory motion and travels in a series of waves. If conditions are right, condensation of water vapor occurs on the rising portion of the wave, and re-evaporation occurs on the falling portion of the wave (see Figure 13). As a result, a lenticular cloud may appear to remain stationary in the sky for a period of several hours. Lenticular clouds are not uncommon in mountainous areas, where the wind

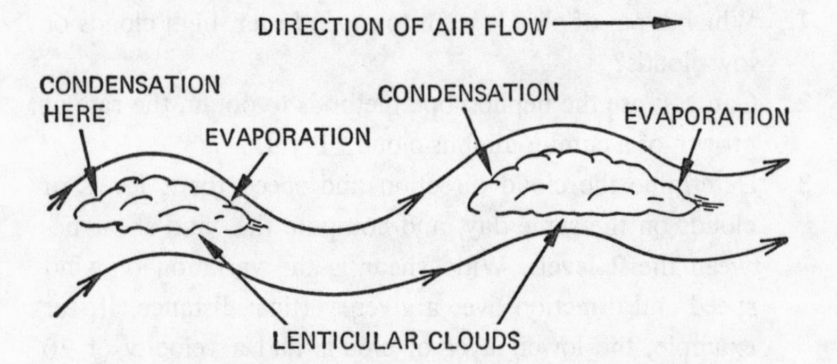

Fig. 13. Diagram Showing Formation of Lenticular Clouds

currents over the mountaintops are set into wavy motion. The clouds have a distinctive appearance, resembling flat wafers or thin lenses. Frequently, several may be seen at one time, extending in a line leeward from the mountain range, or even stacked vertically, one above the other.

INVESTIGATION 6. **(M)**

Measuring Total Sky Cover

MATERIALS NEEDED:

Single-lens reflex camera, such as Pentax, Yashica, or Minolta
Silver Christmas tree ornament, 3 inches in diameter
Wooden dowel, ½ inch by 12 inches
Model-airplane cement
Plywood, 9 inches by 9 inches by ¾ inch
4 wooden posts, 3 feet long
24 feet of sturdy wire
Large washer, 2-inch diameter

Most people are fairly oblivious of the sky's cloud cover on any particular day, except when it results in precipitation that interferes with planned activities. But, of course, the amount and frequency of cloudiness are directly related to the reverse consideration—the number of hours of sunshine and its intensity, a matter of great importance to local chambers of commerce in their efforts to attract visitors and permanent residents. It is fashionable to quote the number of days of sunshine per year at a tourist mecca. One city in the western United States advertises its location as "Where the Sun spends the Winter."

On a worldwide basis, there is variation in total annual sunshine from less than 40 per cent of maximum possible radiation to nearly 100 per cent. Yuma, Arizona, for example, averages 89 per cent of the maximum possible yearly sunshine. (According to the National Geographic Society, Yuma receives more than 4,000 hours of sunshine, twice as much as Seattle, Washington.) At the other extreme Zaïre (formerly the Belgian Congo), a tropical rain forest, averages only 42 per cent of possible annual radiation.

In your own community, perhaps the only observation you've made so far is that the cloudier the day, the less sunshine—true, of course, but can you determine just *how* cloudy the day is? One way of calculating the amount of cloudiness involves photographing the sky cover and measuring the photograph against a grid from which you can calculate the per cent of sky covered by clouds at that particular time. Such a record can be accumulated over a period of time to provide you with the information to answer some interesting questions about cloud formation in your area.

First, a method for photographing cloud cover must be

devised. Obviously, pointing a camera at the sky and snapping a picture will not be satisfactory, since the result of such a procedure will include only a small portion of the area. What we need is a way to *reflect* the total sky cover and take a photo of that reflection.

An inexpensive apparatus for taking sky-cover photographs can be put together from fairly simple materials. You will need a reliable camera, though not necessarily an expensive one. A fixed-focus camera will not be suitable because you will be using the camera to take close-up pictures of a spherical, silvered ball that reflects all portions of the sky simultaneously. This shiny surface poses special problems, and you will need a camera with an adjustable focus, such as a single-lens reflex, to experiment with speed and shutter settings that result in a clear picture.

A silver Christmas tree ornament, spherical, and about 3 inches in diameter, will serve as the reflecting surface. Remove the hanger wire from the ornament and insert a small dowel, approximately ½ inch in diameter and 12 inches long, into the opening, and glue it firmly to the ball with model airplane glue. Screw the other end into a ringstand base or tripod from a chemistry laboratory, or make a small plywood base for the dowel. When in use, the base will rest on level ground, with the camera mounted directly above the spherical ball. A diagram of the setup is shown in Figure 14.

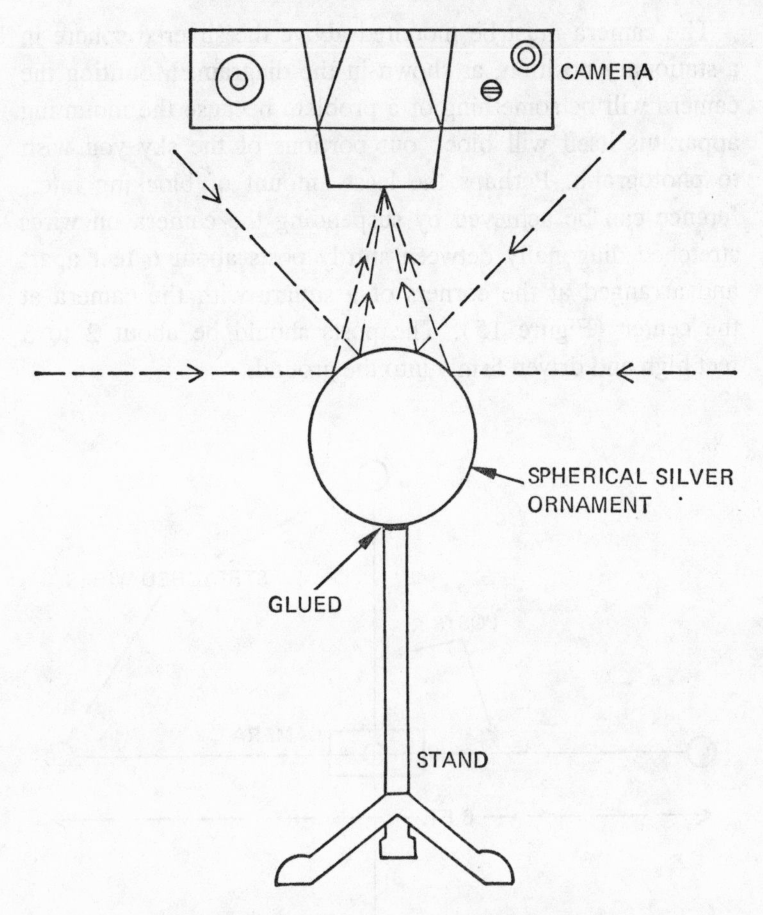

Fig. 14. Setup for Taking All Sky Photographs

The camera must be mounted above the silvered sphere in a stationary position, as shown in the diagram. Mounting the camera will be something of a problem because the mounting apparatus itself will block out portions of the sky you wish to photograph. Perhaps the least amount of blocking interference can be achieved by suspending the camera on wires stretched diagonally between sturdy posts about 6 feet apart and arranged at the corners of a square with the camera at the center (Figure 15). The posts should be about 2 to 3 feet high and driven firmly into the ground.

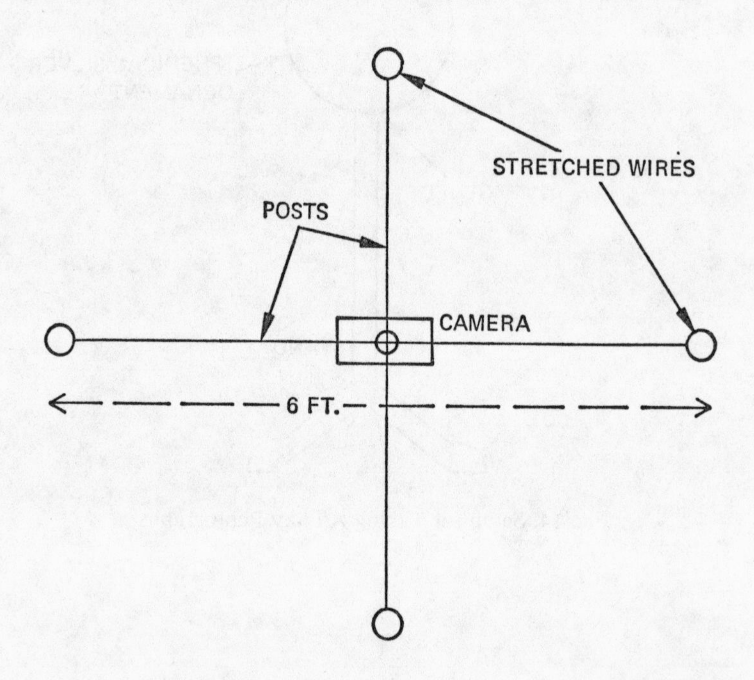

Fig. 15. Diagram Showing a Possible Method of Mounting Camera Above Silver Sphere for All Sky Photographs

If your camera is of the type that has a screw in the back for flash attachments, the camera may be fastened to the support wires by the use of a small washer with a ½-inch hole as illustrated in Figure 16. With no means of screwing

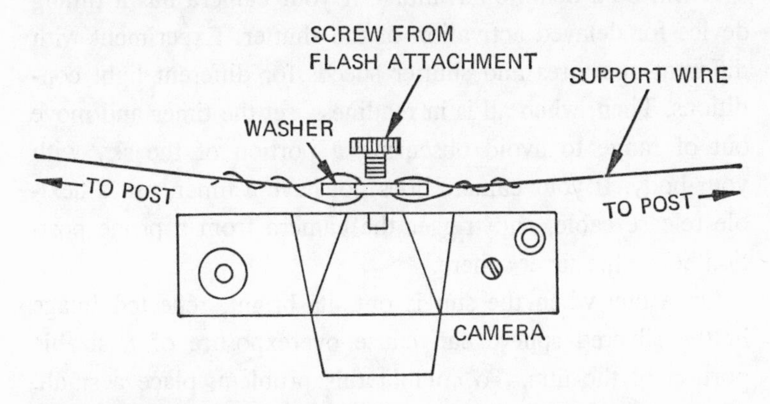

Fig. 16. Diagram Showing Method of Attaching Camera to Support Wires

the camera to the wires, devise a simple wire harness that will cradle the camera safely and that can be hooked to the support wires.

The height of the camera above the silvered sphere should be fixed to take advantage of the best possible viewing and the least amount of blocking of the sky view. A typical single-lens reflex camera mounted 18 inches above the silvered sphere will obscure a visual angle of about 30 degrees. Because of the image formation by the convex hemispherical surface, however, the actual area blocked out by the camera is less than 1⁄100 of the total sky area being photographed, an acceptable loss considering the objectives of the experiment.

To obtain a record of the variations in total sky cover during the day, plan to take photographs at regular intervals throughout the day, say, every 30 minutes from sunrise to sunset.

It will be a definite advantage if your camera has a timing device for delayed activation of the shutter. Experiment with different apertures and shutter speeds for different light conditions. Then, when all is in readiness, set the timer and move out of range to avoid obscuring a portion of the sky with your body. If your camera does not have a timer, use a flexible release cable, and trigger the camera from a prone position below the silver sphere.

On a day when the sun is out, its bright, reflected image in the silvered sphere can cause overexposure of a sizable portion of the film. To combat this problem, place a small, circular piece of black cloth (about ½ inch in diameter) directly on the sphere to blot out the sun's image immediately before snapping the picture. This, of course, will eliminate a portion of the sky being photographed, but is not a serious consequence since no clouds would appear on the reflecting surface in that area anyway, having been blotted out by the sun's brilliant reflection.

Be sure to note the exact time of each photograph. For analysis of the photographs later, you may want to devise a circular grid to enable you to obtain a percentage value of sky cover in order to express your results in quantitative terms.

Such a grid can be made as follows: Select an all-sky photograph made with your camera set up as described above. For best results when photographing, see that the image of the sky as viewed in the spherical ball occupies the largest possible fraction of total print.

Cover the print with a piece of onion skin paper, which is semitransparent. Trace a circle in pencil on the onion skin paper the same size as the sky image on the photograph.

Measure the diameter of the circle. As an example, suppose this turns out to be 3.5 centimeters. Since the area of the whole circle represents 100 per cent of the sky, half the area of the circle would represent 50 per cent of the sky. To find what diameter is needed to make a circle representing 50 per cent of the sky, use the relation $A=\pi/4\ d^2$. Let A_1=the area of the large circle, d_1 its diameter. Let A_2=the area of the small circle, d_2 its diameter. To obtain d_2, set up the proportion

$\frac{A_1}{A_2}=\frac{\pi/4\ d_1^2}{\pi/4\ d_2^2}$. This reduces to $\frac{A_1}{A_2}=\frac{d_1^2}{d_2^2}$

Solve the proportion for d_2.

$$d_2=\sqrt{\frac{A_2\, d_1^2}{A_1}}$$

If the measured diameter of the large circle was 3.5 centimeters, the solution would be:

$$d_2=\sqrt{1/2\ (3.5\text{ cm})^2}=\frac{1}{\sqrt{2}}\times 3.5\text{ cm}$$
$$=2.5\text{ cm}$$

Draw a circle of 2.5 centimeters diameter inside the circle of 3.5 centimeters diameter. Using the relation above, find the diameter, d_3, of a third circle, which has 50 per cent of the area of the second circle. Draw a circle of this new diameter inside the second circle. If space permits, repeat the process again by drawing an additional smaller circle on the onion skin paper (Figure 17). Divide the circle into sixteenths by radial lines as shown in the figure.

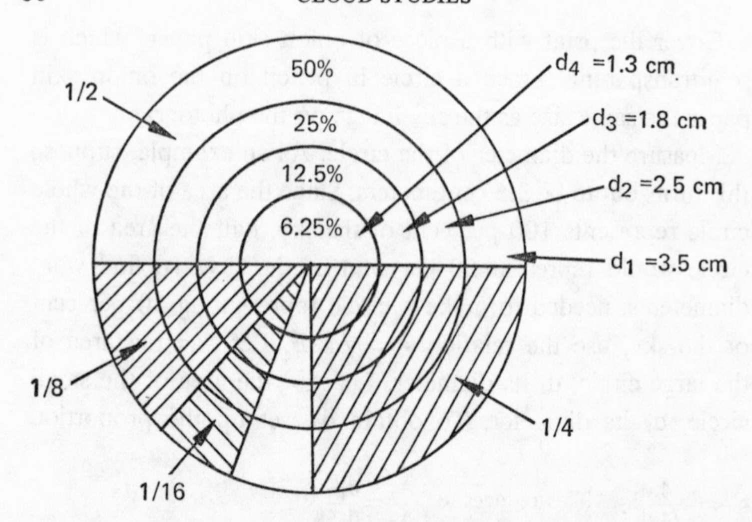

Fig. 17. Circular Grid Useful in Estimating Per Cent of Sky Cover

After all lines are completed, make a transparent copy of the circular grid using Thermofax or some other duplicating process. (Thermofax machines are commonly available in libraries and offices.)

For use, lay the transparent circular grid over the sky image to be analyzed. By counting the areas in which clouds are found, you can get a reliable estimate of the per cent of sky covered by clouds.

Some questions that come to mind when considering sky cover are the following:

1. Why is quantitative determination of sky cover needed? How is total sky cover related to the intensity of solar radiation received at the ground? There is an instrument, called a pyrheliograph, which keeps a continuous record of solar radiation received on a horizontal surface, in calories per square centimeter per minute. Weather sta-

tions frequently have such instruments and keep records of this kind. Can this record be correlated with per cent of total sky cover?

2. How do different clouds affect the per cent of total sky cover? Which types of clouds tend to obscure the sky the most? Which affect it the least?
3. How rapidly does sky cover change, for example, from clear to scattered to broken to overcast? The usual designations for these terms are:

   Clear: 0 to .1 coverage
   Scattered: .1 to .4 coverage
   Broken: .5 to .8 coverage
   Overcast: More than .9 coverage

4. Does cloud formation in your area follow a regular day-to-day pattern, for example, clear in the morning, broken at noon, overcast in the afternoon, etc.?
5. Can you identify seasonal variations in sky cover? What do they mean in terms of forecasts of rainfall or maximum temperatures achieved during the daytime?

Chapter 3
ATMOSPHERIC MOISTURE

The presence of water in the atmosphere is most obvious to us when it appears in one of the many forms of condensation and precipitation—rain, snow, fog, or dew, for example. And yet, water, whether in its solid, liquid, or gaseous state, is indirectly involved in many other weather phenomena not immediately associated with atmospheric moisture.

The wind seems to be one meteorological force free of the influence of atmospheric moisture, but closer study reveals that the two, in fact, are closely interrelated. Water has an indirect influence on the effects of wind because the amount of water vapor carried by moving air is directly related to its cooling effect. Thus, the drier the air, the more cooling effect it has for any given wind speed and temperature.

Although it is not always apparent, radiation is another example of an atmospheric event affected by moisture. Actually, radiation is affected by water in at least 2 ways. First, the water vapor in the air absorbs long-wavelength radiation from the sun (infrared rays) and slightly reduces the amount of solar radiation that reaches the ground. And, of course, clouds themselves, consisting of water droplets or ice crystals, effectively control the amount of sunshine that reaches the earth. Second, radiation from the earth is strongly absorbed by the water vapor in the air, thus raising the temperature of the atmosphere. This trapping phenomenon is an important factor in man's existence and, as we noted before, it is called the "greenhouse effect."

Until condensation and precipitation occur, the water in the atmosphere is invisible. Consequently, we tend to forget that it is a vital component of the air that surrounds us at all times. We breathe water vapor into our lungs with every breath and add to the vapor in the air when we exhale.

The water vapor in the air is called humidity, and there are several ways of expressing it: relative humidity, specific humidity, mixing ratio, and absolute humidity, also called vapor density. Relative humidity is the term commonly used on the radio and in newspapers to report atmospheric conditions to the non-meteorologist. Because the layman is familiar with the expression "relative" humidity and has come to associate it with his personal feelings of comfort and discomfort, he probably finds it the easiest term to understand of all the possible methods for expressing the level of humidity in the air. To the meteorologist, however, relative humidity is not a particularly useful concept because of an inherent problem. Relative humidity changes markedly with small changes in air temperature, even without the addition or subtraction of any moisture. For this reason, the meteorologist usually prefers to use specific humidity or mixing ratio.

Specific humidity is the actual number of grams of water vapor per kilogram of *moist* air and does not change with temperature. Mixing ratio is the number of grams of water vapor per kilogram of dry air and is numerically very close to the specific humidity. If, for example, the specific humidity on a given day is 5.00 grams per kilogram, the mixing ratio on the same day is 5.03 grams per kilogram. To understand how these 2 measurements are related, consider the formulas for calculating specific humidity and mixing ratio:

let m_v=the mass of water vapor in the air

m_d=the mass of dry air

$$\text{specific humidity} = \frac{m_v}{m_v + m_d} = \frac{5 \text{ grams}}{5 \text{ grams} + 995 \text{ grams}} = 5 \text{ grams per kilogram}$$

$$\text{Mixing ratio} = \frac{m_v}{m_d} = \frac{5 \text{ grams}}{995 \text{ grams}} = 5.03 \text{ grams per kilogram}$$

Depending on the air temperature, the specific humidity might vary from a few tenths of a gram per kilogram to more than 20 grams per kilogram of moist air, and the corresponding mixing ratios will always be slightly larger.

Absolute humidity may be defined as the number of grams of water vapor per cubic meter of moist air. While absolute humidity is an expression of weight, specific humidity—measured in grams per kilogram—refers to the mass of water vapor per unit mass of moist air.

Relative humidity, on the other hand, is always expressed as a per cent value and is the ratio of the amount of water vapor in the air to the amount the air could hold at a given temperature. The capacity of the air to hold water vapor increases as the temperature increases. A rise in the temperature of the air, therefore, results in a decrease in the relative humidity, even though there is no actual change in the amount of water vapor present in the air. With this very brief introduction to the subject of humidity, you can begin your own studies.

Any systematic approach to analysis of atmospheric moisture must begin with a reliable means of measuring humidity. Five ingenious measuring devices are described in this chapter. Perhaps you will find the construction of the dew cell and the device for measuring humidity by infrared absorption most challenging. These instruments call for access to specialized equipment. On the other hand, the sling psychrometer or the hair hygrometer are simple and perfectly adequate devices that can open for you broad areas of investigation into local variations in atmospheric humidity.

INVESTIGATION 7. (E)

Finding Local Variations in Relative Humidity

MATERIALS NEEDED:

Materials for sling psychrometer:
2 Fahrenheit thermometers (−10° to +110° F range)
Plywood block, 10 inches by 4 inches by ½ inch
6 small screws, ½ inch long
Wooden dowel, 1 inch diameter, 6 inches long
2 small washers
1 screw, 2 inches long
Small square of cheesecloth, 2 inches by 2 inches

If you constructed the sling psychrometer described earlier (see Investigation 4), you have at hand a most useful device for measuring relative humidity. And if you have access to a varied topography—hills, a body of water, perhaps—you have all the raw materials for an interesting investigation in relative humidity variations. Relative humidity can be measured accurately by use of the sling psychrometer used in the experiment on measuring cloud height. The procedure is identical to that used to determine dew point. Find the dry-bulb and wet-bulb temperatures, compute the difference between them, and refer to the section of the psychrometric tables dealing with relative humidity. Read downward on the left side to the dry-bulb temperature, and horizontally to the computed difference value. The relative humidity in per cent is found at that point.

To study relative humidity systematically, obtain a map of the area you wish to study—or draw your own simple map—and mark on it the specific locations you can reach conveniently on a regular basis. Select a variety of sites such as a hilltop, the shore of a small body of water, a wooded area, a grassy field, a playground, an asphalt parking lot—whatever locations offer as much variety in topography, exposure, and nearness to water as possible within a workable area.

Make your observations using the sling psychrometer on a regular basis, such as at 8:00 A.M., 12:00 noon, 4:00 P.M., and 8:00 P.M., every day in each of the selected locations. Take all readings in the shade to avoid the effects of direct sunlight on the psychrometer. Note the exact time for each observation and repeat the observations at the same times for at least 10 days. At each observation, record the following information on a data sheet: location, time, date, wet-bulb degrees in Fahrenheit, dry-bulb degrees in Fahrenheit, relative humidity per cent, wind direction, wind speed (est. mph), cloud cover (0 to .9), comments (precipitation, etc.).

Wind speed can be estimated accurately enough by using the Beaufort Wind Scale reproduced below.

The Beaufort Scale of Wind Force With Specifications and Velocity Equivalents

Beaufort Number	*General Description*	*Specifications*	*Velocity meters per sec.*	*Velocity miles per hour*
0	Calm	Smoke rises vertically	Under 0.6	Under 1
1	Light air	Wind direction shown by smoke drift but not by vanes	0.6–0.7	1–3
2	Slight	Wind felt on face; leaves rustle; ordinary vane moved by wind	0.8–3.3	4–7
3	Gentle breeze	Leaves and twigs in constant motion; wind extends light flag	3.4–5.2	8–11
4	Moderate breeze	Dust, loose paper, and small branches are moved	5.3–7.4	12–16
5	Fresh breeze	Small trees in leaf begin to sway	7.5–9.8	17–22
6	Strong breeze	Large branches in motion; whistling in wires	9.9–12.4	23–27
7	Moderate gale	Whole trees in motion	12.5–15.2	28–34
8	Fresh gale	Twigs broken off trees; progress generally impeded	15.3–18.2	35–41
9	Strong gale	Slight structural damage occurs; chimney damage	18.3–21.5	42–48
10	Whole gale	Trees uprooted; considerable structural damage	21.6–25.4	49–56
11	Storm	Very rarely experienced; widespread damage	25.5–29.0	57–67
12	Hurricane		Above 29.0	Above 67

When you have obtained data for the period of study, you can tabulate it in a variety of ways for analysis. For example, you might wish to find the average relative humidity at each location at the time of observation. Plotting these data on a graph of relative humidity versus time of day will show the daily changes that occur on a regular basis. Such a graph might look like the one in Figure 18. Or you might want to

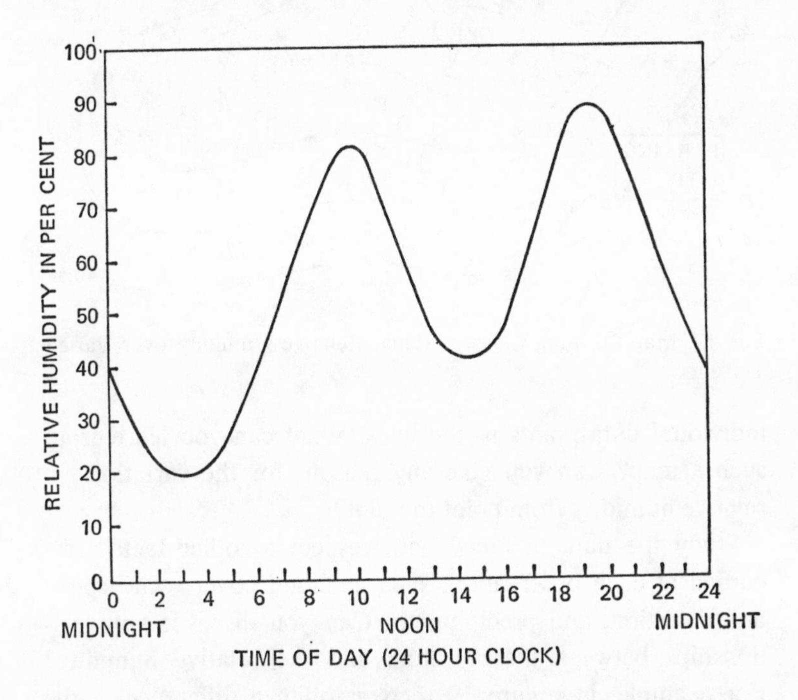

Fig. 18. Graph Showing Possible Daily Fluctuations in Relative Humidity

see how much fluctuation occurs in the relative humidity at a particular location over the period studied.

After you have obtained the average relative humidity at each location, plot the information on a detailed map of the area, putting down the average values obtained at each of the observation points. Draw lines of equal relative humidity on the map, as shown in Figure 19, positioning them so the

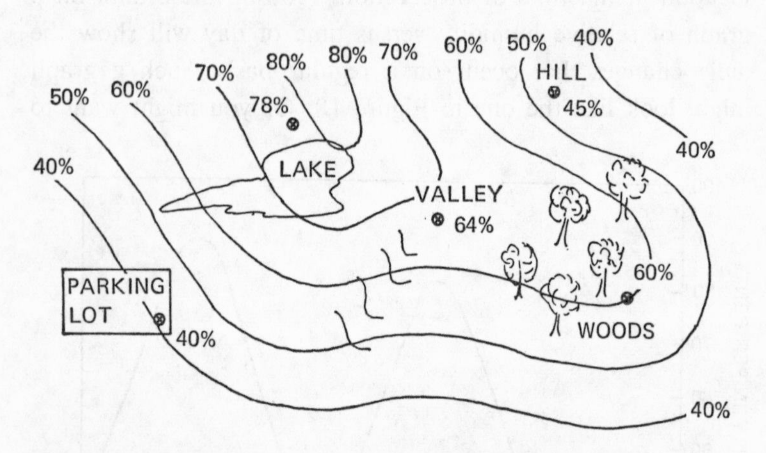

Fig. 19. Map Showing Lines of Equal Relative Humidity over Varied Terrain

individual data points fit the lines. What can you learn from such a map? Can you give any reasons for the variations in relative humidity from point to point?

Study the data obtained with respect to other factors recorded at each observation, such as cloud cover, wind speed and direction, and precipitation. Can you discover any relationships between these factors and the relative humidity? For example, in Figure 19 there is quite a difference in the relative humidity between one end of the lake and the other—from 50 per cent to 80 per cent. By comparing the map with a hypothetical data sheet, it might become apparent that the wind direction at the time of observation was from west to east. The air would thus pick up moisture and become more humid as it passed over the surface of the lake, explaining why the relative humidity is so much higher at the other end of the lake. Through careful co-ordination of all information on the contour maps and data sheets, try to obtain as much information as possible about variations in relative humidity and their causes.

INVESTIGATION 8. (E)

Construction of a Hair Hygrometer

MATERIALS NEEDED:

Cardboard milk carton (1-quart size)
2 small buttons
1 large darning needle, 3 inches long
Small fragment of pocket mirror, ¾ inch by ¾ inch
1 tongue depressor
1 soda straw
Model-airplane glue
1 human hair, 10 inches long
Masking tape
Graph paper, fine grid

The hair hygrometer has been used for many years to measure relative humidity. Although not quite as accurate as the sling psychrometer, it has the advantages of continuous operation and convenience. Hair is used as the working agent because of its property of absorption and release of water vapor with varying humidity conditions. When placed under slight tension, the hair tends to stretch as it absorbs moisture and shrink as it dries out. This mechanical action is translated, in the hair hygrometer, into movement of a light pointer on a scale calibrated to give a percentage reading of the relative humidity. Thus, once the hygrometer is in working condition, readings can be taken at any time.

Begin with a blond human hair about nine or ten inches long. (Blond hair is more sensitive to humidity conditions and therefore works better than brunette or red hair.) Wash the hair in warm, soapy water, rinse thoroughly, and set aside to dry. Washing removes the natural oils, which would otherwise interfere with proper absorption of moisture.

In an empty, rinsed-out cardboard milk carton, cut a small H (about two inches long and one inch wide) in one side near one end and bend the tabs formed downward, as shown in the diagram (Figure 20). Then cut three long rectangular

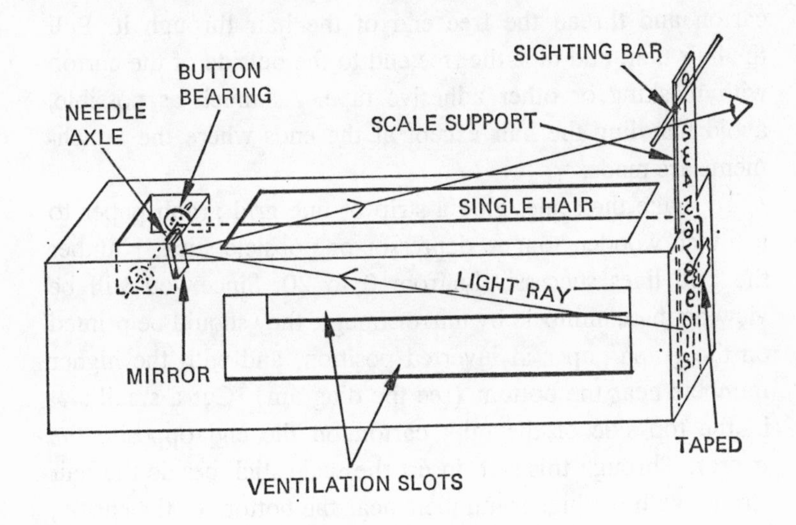

Fig. 20. Simple Hair Hygrometer

openings in the top and 2 sides of the carton, as shown. These slots will allow air to circulate freely through the box.

To each of the tabs, glue a small shirt button in position to form a bearing for the axle, which should be a 3-inch sewing needle or a length of thin, straight wire. Mount the needle in position through the bearings. At the midpoint of the needle, glue a small fragment, about ¾ inch square, of a broken pocket mirror. The mirror, as you might guess, will reflect light from the scale, enabling the viewer to read the humidity measurement directly. The needle is glued to the back of the mirror near one end so that when the mirror is in position, its weight allows it to hang nearly vertically.

Attach one end of the hair to the needle axle near the mirror, using a drop of glue. When the glue is dry, wind the hair 2 or 3 turns around the needle in a direction such that pulling on the hair brings the front surface of the mirror forward. Drill a small hole in the opposite end of the milk carton and thread the free end of the hair through it. Pull the hair taut, and tape the free end to the outside of the carton with masking or other adhesive tape. As much as possible, avoid handling the hair except at the ends where the attachments are made.

To make the scale, glue a strip of fine grid graph paper to a 6-inch wooden tongue depressor or popsicle stick. Number the grid lines successively from 0 to 20. Since you will be viewing these numbers by mirror image, they should be printed on the graph paper in inverted position, and with the higher numbers near the bottom (see the diagram). Cut a small slot in the top side of the milk carton on the end opposite the mirror. Through this slot, insert the scale stick beside the hair strand, with the higher numbers near the bottom of the carton. Tape the scale in position with masking tape. At the top of the scale stick, tape a 2-inch length of straightened paper clip horizontally to form a sighting bar to be used in making your observations.

To adjust the mirror of the hygrometer, detach the taped end of the hair temporarily. Sight over the sighting bar toward the point on the mirror where it is glued to the axle. Pull on the hair until the number 10 on the scale comes into view and is on the line of sight. Retape the free end of the hair with the mirror in that position. Your hair hygrometer is now ready for calibration, and the best procedure involves using the sling psychrometer.

Place the hygrometer on a level surface in the bathroom. Turn on the hot shower and let it run until the windows and mirror cloud up. Allow the hair hygrometer to come to equilibrium in this humid atmosphere, which will take about 15 minutes as the hair absorbs moisture, and stretches. At the end of that time, read the scale by sighting over the sighting bar toward the mirror, and record the reading. Next, obtain the relative humidity from a sling psychrometer reading, and write this figure next to the hygrometer reading you have recorded.

Using the sling psychrometer, obtain the relative humidity in 3 or 4 other places, such as outdoors and inside a heated house. Each time, allow the hair hygrometer to come to equilibrium, and read the scale. Record the scale reading and the relative humidity obtained from the sling psychrometer side by side on the calibration chart. After you have 4 or 5 readings spread out over the entire length of the scale, fill in the gaps on the calibration chart by assigning appropriate relative humidities for the remaining numbers.

A final check of the accuracy of the hair hygrometer should be made by comparing it with the sling psychrometer under a variety of conditions and over a period of time. From this comparison, you can devise a correction scale that can then be applied to your hair hygrometer to give more accurate readings.

A sample set of data obtained in such a calibration check might look like this:

	Date	Time	Weather Description	Hygrometer Reading (per cent)	Psychrometer Reading (per cent)	Deviation
1	5-15	5:30 P.M.	cloudy; just rained	85	88	− 3
2	5-15	10:00 P.M.	raining	100	94	+ 6
3	5-16	5:00 P.M.	sunny	25	20	+ 5
4	5-16	10:00 P.M.	clear	15	19	− 4
5	5-17	7:30 A.M.	clear	10	7	+ 3
6	5-17	12:30 P.M.	clear	10	15	− 5
7	5-18	11:30 A.M.	mostly clear	10	13	− 3
8	5-18	9:15 P.M.	clear	15	16	− 1
9	5-19	8:30 A.M.	clear-cloudy	55	57	− 2
10	5-19	4:30 P.M.	cloudy	85	75	+10
11	5-20	10:00 A.M.	cloudy	35	34	+ 1
12	5-20	4:00 P.M.	clear	15	12	+ 3
13	5-21	8:30 A.M.	clear-few clouds	20	22	− 2
14	5-21	3:00 P.M.	clear	15	no measurement	—
15	5-22	9:30 A.M.	clear	10	16	− 6
16	5-22	1:30 P.M.	clear	15	19	− 4
17	5-23	10:00 A.M.	clear	10	13	− 3
18	5-23	10:00 P.M.	sprinkling	95	94	+ 1
						AVE.: −0.22%

Plotting the two sets of readings gives the graph shown in Figure 21.

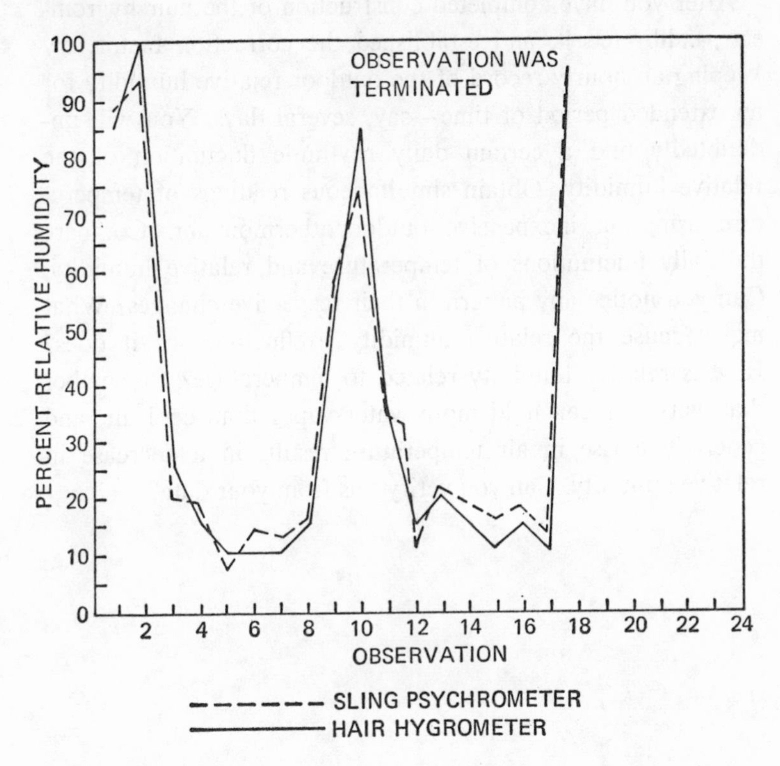

Fig. 21. Comparison Between Hair Hygrometer and Sling Psychrometer Readings

Such a comparison of data obtained from the hair hygrometer and the sling psychrometer discloses small deviations in most samples. The average deviation in this case, for example, is minus 0.22 per cent, which means the hair hygrometer averages about 0.22 per cent low in its readings, as compared with the sling psychrometer. The correction to be applied, then, to subsequent readings of the hygrometer is plus 0.22 per cent. This is such a small correction, however, that it can be ignored in this example.

After you have completed construction of the hair hygrometer, calibrated it, and established the correction factor, try keeping an hourly record of the outdoor relative humidity for an extended period of time—say, several days. You will undoubtedly find a certain daily rhythmic fluctuation of the relative humidity. Obtain simultaneous readings of temperature using an inexpensive outdoor thermometer. Compare the daily fluctuations of temperature and relative humidity. Can you notice any pattern in their respective changes? What might cause the relative humidity to fluctuate as it does? How is relative humidity related to temperature? Remember that warm air can hold more water vapor than cool air, and generally a rise in air temperature results in a decrease in relative humidity. Can you verify this from your data?

INVESTIGATION 9. (D)

The Dew Cell—
a Direct-Reading Hygrometer

MATERIALS NEEDED:

Fahrenheit thermometer, –10° F to +220° F range
One-hole rubber stopper to fit thermometer, size No. 5 or 6
Tube of silicone rubber RTV731
Silver wire, about 2 feet, 24–28 gauge
Lithium chloride solution (saturated)
25-volt filament transformer
2 200-ohm electrical resistors, 10 watts each
2 3/16-inch bolts, 2 inches long, with nuts
Acetone or alcohol
Household cement (e.g., Glyptal)
Distilled water

A somewhat more sophisticated method of measuring humidity involves the dew cell. The dew-point temperature, as noted earlier, is widely used as a measure of atmospheric humidity and is defined as the temperature to which a parcel of air must be cooled in order just to begin to condense the water vapor in the air.

The dew cell will give continuous readings of the dew-point temperature of the air to within ±1.0° F. In addition, unlike the wet- and dry-bulb psychrometer, the dew cell provides a direct measurement of the dew point, making the use of psychrometric tables unnecessary, and it can be used at temperatures below freezing because it is not affected by water freezing on the wick. More importantly, however, the cell does not add or subtract moisture from the air and therefore can be used to measure the dew point of small volumes of air.

A schematic diagram of a dew cell is shown in Figure 22. Lithium chloride (LiCl), the active agent, is painted onto the bulb of a mercury thermometer. Since lithium chloride is hygroscopic, or moisture-absorbing, it forms an electrically conducting solution as it absorbs water vapor from the atmosphere. Alternating current passing through silver wire electrodes heats the bulb and raises the temperature of the LiCl and the thermometer. Increasing temperature tends to drive off water from the lithium chloride solution, thereby decreasing the conductivity and heating effect of the solution. As a result, the thermometer registers a drop in temperature. Eventually, an equilibrium temperature is established at which the rate of absorption of water vapor from the atmosphere is equal to the rate at which water is driven off from the solution. This equilibrium temperature is a measure of the dew point of the air, and will automatically respond to changes in the water vapor content of the atmosphere.

When you have accumulated the necessary materials, begin assemblage of the dew cell with the thermometer attachment.

Run the 2 contact bolts through the rubber stopper, as shown in Figure 22, and push the stopper onto the thermometer until the entire bulb is exposed. Clean the thermometer bulb thoroughly with acetone or alcohol. Paint strips of RTV731 or other silicone rubber the full length of the thermometer bulb on opposite sides of the bulb to support the windings of wire.

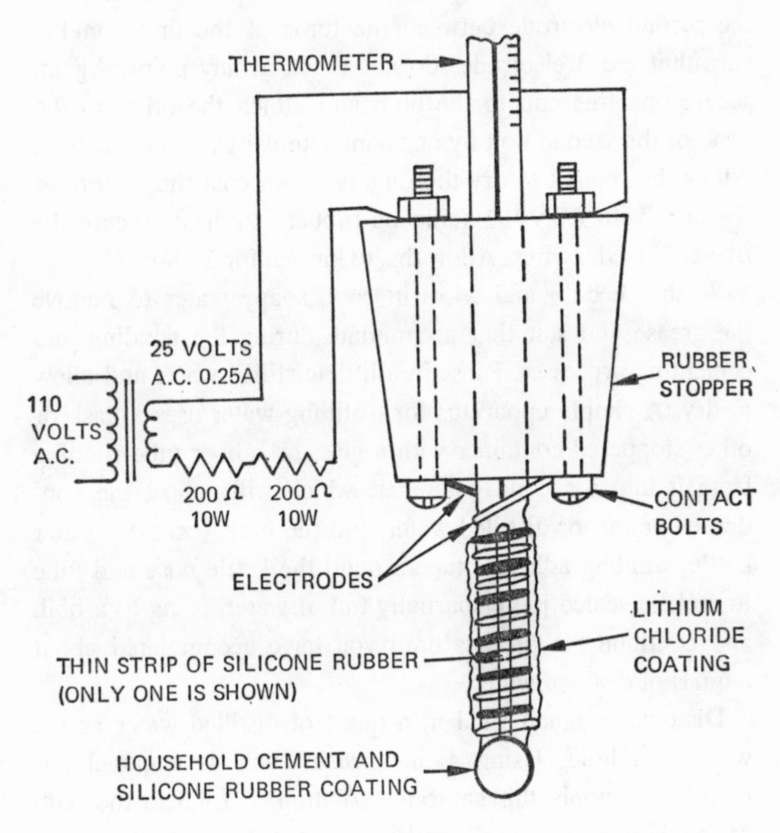

Fig. 22. Dew Cell

(Silicone rubber can be purchased in hardware stores.) Allow it to cure for 24 hours.

The silver wire will become the electrodes of the apparatus. Snip the length in half and wind the first electrode around the bulb with 1/16-inch spacings between turns. Secure the bottom free end to the bulb with a drop of fast-drying household cement. Attach the other end to the head of one bolt by one complete twist around the bolt. Cut off any excess wire. Wind the second electrode between the turns of the first, making sure that the 2 electrodes do not touch at any point. Again, secure one free end to the bulb and attach the other to the head of the second bolt by one complete twist around the bolt. Allow the cement to dry thoroughly. Then coat the bottom of the bulb with RTV731 (silicone rubber) to firmly secure the free ends of the wires. Allow this to harden for 24 hours.

Wash the bulb and wires in cool, soapy water to remove the grease and salt that accumulate during the winding and cementing processes. Rinse in a little distilled water and allow to dry. A simple apparatus for distilling water uses a flask or other stoppered container with a glass or rubber tube running from it into a second container, which will collect the condensed vapor, or distilled water. Fill the flask (or use a water kettle, winding adhesive tape around the kettle nose and tube to make a sealed joint) partially full of water, bring to a boil, and continue the process until you have accumulated about a quart of distilled water.

Dissolve as much LiCl in a quart of distilled water as the water will hold. Using a fine brush, such as a camel-hair paintbrush, apply this saturated solution of LiCl to the bulb of the thermometer until you have a uniformly thin film. The

LiCl should coat the electrode wires completely. Complete the circuit by attaching the electrical resistors as shown in Figure 22, and apply voltage. The temperature will rise slowly at first as water is driven from the dilute LiCl solution, and then will abruptly increase to over 220° F. *Caution: Do not let the rise in temperature exceed the maximum capacity of the thermometer. Disconnect the current, if necessary, to prevent this from happening.* After the initial rise, the temperature will decrease until it reaches the equilibrium temperature.

The dew cell functions best in a still atmosphere, but it can be made to operate in slight wind or rain by enclosing the thermometer bulb in a 2-inch length of protective covering such as metal, glass, or plastic tubing. Start the dew cell in still air without using a shield. On initial heating, large amounts of water are given off from the dilute LiCl solution, and these are most rapidly dispelled when there is no shield to interfere.

The table below shows a conversion scale for converting the thermometer reading to dew-point temperature. You may wish to prepare a piece of paper with the dew cell temperatures placed horizontally with the appropriate dew-point temperatures and paste it against the thermometer when it is mounted in position for use. Figure 23 shows the dew cell in operating position.

There are conditions under which the dew cell cannot achieve a dew-point measurement, primarily when the equilibrium temperature, if it were attainable, is below the ambient (surrounding air) temperature. This situation will exist when the relative humidity is below about 15 per cent. The condition can be alleviated by using a lithium bromide (LiBr) solution to coat the thermometer bulb in the dew cell. LiBr

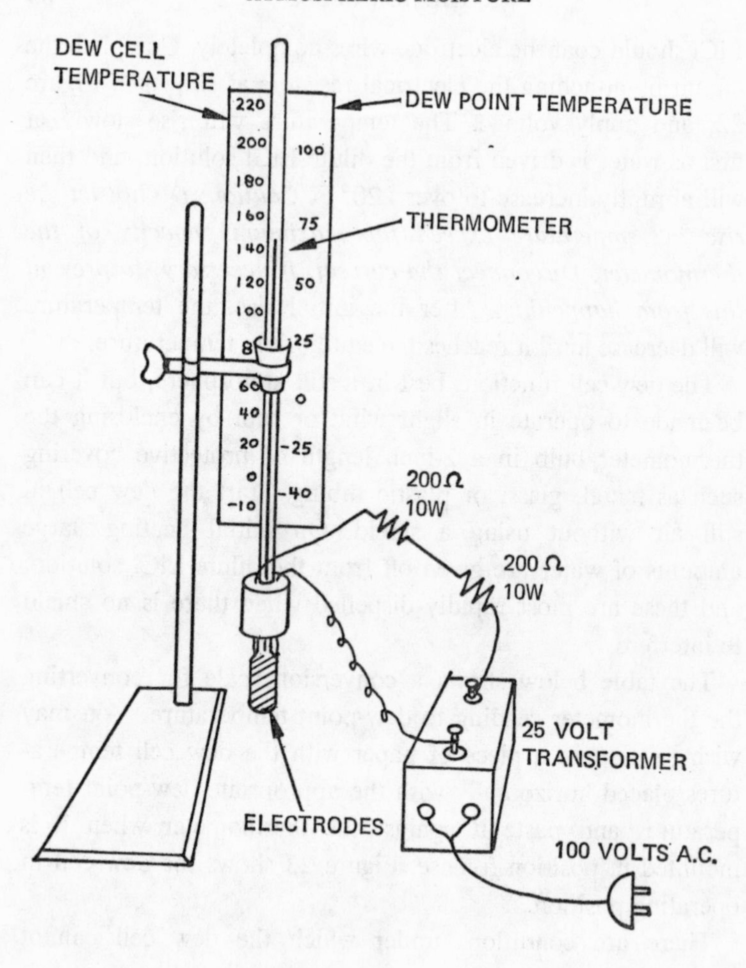

Fig. 23. Dew-Cell Setup

has greater hygroscopicity than LiCl, and by substituting LiBr in the dew cell, the attainable equilibrium temperatures correspond to relative humidities below 8 per cent. The table also gives the conversion scale for converting LiBr thermometer readings to the corresponding dew-point temperatures.

Dew-point Temperature ° F	*LiCl Dew-Cell Temperature ° F*	*LiBr Dew-Cell Temperature ° F*
– 40	– 3.1	6.5
– 35	3.9	13.2
– 30	11.1	20.1
– 25	18.0	27.1
– 20	24.9	34.2
– 15	32.0	41.5
– 10	38.8	48.9
– 5	45.7	56.7
0	52.5	64.6
5	59.3	72.9
10	66.0	81.5
15	72.8	91.0
20	79.5	103.5
25	86.2	111.9
30	92.8	118.8
35	99.4	125.8
40	105.8	132.6
45	112.4	139.8
50	119.3	146.8
55	126.5	154.0
60	133.8	161.2
65	141.2	168.4
70	148.5	176.0
75	155.9	183.4
80	163.3	190.8
85	171.1	198.3
90	179.3	206.6
95	188.0	214.2
100	196.0	222.3

Make periodic comparisons of dew-point values obtained from the dew cell and from the sling psychrometer. How closely do they agree? Which is probably the more accurate?

Forecasting dew-point temperature is an important aspect of the meteorologist's work because of the need for predicting the advent of fog, particularly at an airport. Conditions are most favorable for fog formation after a period of rain and late afternoon clearing of the sky, with subsequent radiational cooling of the air near the ground.

Try to forecast the time of fog occurrence when these conditions are present. You will need to keep a continuous record of temperature and dew point at intervals of every thirty minutes, or every hour, beginning in the early afternoon of a day when the necessary conditions are likely to materialize. After you have obtained several readings, attempt to determine the rates of decrease in temperature and dew point by plotting both sets of information (temperature versus time of day) on a graph. From this graph, make an educated guess as to the time fog will probably occur.

A sample graph might look like this (Figure 24):

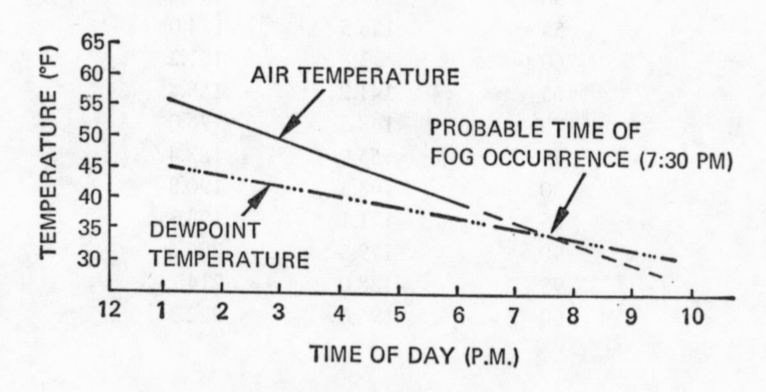

Fig. 24. Graph for Predicting Occurrence of Fog

INVESTIGATION 10. (D)

Direct Measurement of Absolute Humidity

MATERIALS NEEDED:

Analytical balance capable of measuring to hundredths of a gram
Liter flask with a solid stopper
Small porcelain crucible
Hygroscopic salt (NaCl, $CaCl_2$, or LiCl)

We have already defined absolute humidity, sometimes called vapor density, as the number of grams of water vapor per cubic meter of moist air. (Specific humidity, on the other hand, is measured in grams per kilogram and refers to the *mass* of water vapor per unit *mass* of moist air.)

Saturated air—air that contains all the water vapor it can hold—at 75° F at sea level contains approximately 6.25 gram of water vapor per cubic meter. This value is obtained by using the U. S. Weather Bureau psychrometric tables, which give saturated vapor pressures at various temperatures, and by applying the following equation:

$$\text{Absolute humidity} = 216.5\frac{e}{p},$$

where e is saturation vapor pressure in inches of mercury at a temperature of 75° F, p is sea level atmospheric pressure in inches, and 216.5 is a constant of proportionality to convert to units of grams per cubic meter. The result is:

$$\begin{aligned}\text{Absolute humidity} &= (216.5)\,\frac{(0.866\ \text{inch})}{(30.00\ \text{inches})}\\ &= 6.25\ \text{grams per cubic meter}\end{aligned}$$

If the air at 75° F were only 50 per cent saturated (or, in other words, had a relative humidity of 50 per cent), the absolute humidity would be 3.12 grams per cubic meter.

The basis for an experiment in measuring absolute humidity is the hygroscopic nature of certain salts, such as sodium chloride (table salt), calcium chloride, or lithium chloride. Since these salts tend to absorb moisture directly from the atmosphere and, in doing so, gain weight, it follows that the difference in weight before and after the salt has been exposed to air is a direct indication of humidity.

Prepare a quantity of the salt by drying it in a shallow pan in an oven at 200° F overnight. The heating will drive off most of the water initially contained in the salt. Weigh a quantity of the dried salt (about ten grams) in a small porcelain crucible, to the nearest hundredth of a gram, if possible. Work rapidly to minimize water absorption during the weighing process. Place the weighed salt and crucible inside a liter flask that has been opened to the air, the absolute humidity of which is to be determined. Close the flask with the rubber stopper and let it sit for 24 hours.

Reweigh, to the nearest hundredth of a gram, the salt and crucible. Again, work rapidly to minimize water absorption during the weighing. The resulting gain in weight of the salt represents the water vapor absorbed from one liter of air. Multiplying this value by 1,000 (because there are 1,000 liters in one cubic meter) gives the absolute humidity in grams per cubic meter. For more accurate results, repeat this experiment 10 times, or use 10 separate samples of the hygroscopic salt, and average the results. You might also use a control (for example, equal quantities of a non-hygroscopic material, such as fine sand) to provide a standard against which the weight gains can be compared.

Experiment using samples of air from a variety of locations, such as near lakes, under a clump of trees, or over an asphalt parking lot, and compare the results. Obtain air samples under different weather conditions, such as clear and foggy days, and during the day and at night. Again, compare the results.

When you have a good range of measurements, see if you can answer some questions about this technique for measuring absolute humidity:

1. What are possible sources of error? What are reasonable expectations? How can errors be minimized?
2. Do hygroscopic salts absorb other gases besides water vapor, such as carbon dioxide and oxygen, which would tend to negate your results? A chemistry book will help you here.
3. Which hygroscopic salts seem to give the best results

in this experiment? Could these differences be due to variations in absorption rate or capacity?

4. What is the condition of the air in the liter flask after the 24-hour period? Is it completely dry?
5. Is there any value in your procedure for meteorological purposes? Can you suggest any refinements in the technique?

INVESTIGATION 11. (D)

Measuring Relative Humidity by Infrared Absorption

MATERIALS NEEDED:

1 metal container, 1-quart capacity, painted dull black on outside
1 metal container, 1-quart capacity, polished shiny on outside
1 large flashlight

Test materials 50 centimeters by 50 centimeters square

a. Polished aluminum or aluminum foil
b. Glass
c. Plexiglass
d. Polyethylene
e. Heavy fabric
f. Sheer nylon or silk fabric sheets

20 6-inch lengths of uninsulated copper wire, gauge No. 20
20 6-inch lengths of uninsulated constantan wire, gauge No. 20
1 bakelite plate, ¼ inch by 6 inches by 3 inches
1 microammeter .001 ampere to .1 ampere range
2 Fahrenheit thermometers, 10° F to 220° F range
2 No. 6 rubber stoppers, one hole
1 cylindrical stovepipe, 6-inch diameter and 3 feet long

Radiant energy coming from the sun includes wavelengths ranging from very short X-rays to long radio waves. As these wavelengths pass through the atmosphere, they are absorbed and attenuated in varying degrees depending on the wavelength of the radiation and on the particular components of the atmosphere. Most of the wavelengths shorter than 3,000 Angstrom units (0.3 micron), including ultraviolet radiation and X-rays, are absorbed effectively by oxygen in the atmosphere.

Visible light of 3,900 to 7,600 Angstroms (0.39 to 0.76 micron) generally penetrates the clear atmosphere unhindered, and is absorbed by and warms the earth's surface. As the earth heats up, it begins to reradiate energy outward into the atmosphere. This radiant heat consists of very long wavelengths, from about 7,600 to 300,000 Angstroms (0.76 micron to 30 microns), and is in the infrared portion of the spectrum. The atmospheric components that absorb the infrared wavelengths are water vapor, which absorbs wavelengths in the range of about 5.3 to 7.7 microns and beyond 20 microns; ozone, 9.4 to 9.8 microns; and carbon dioxide, 13.1 to 16.9 microns. Since ozone and carbon dioxide have relatively low absorptivities, their effects can be ignored for many atmospheric radiation studies, leaving water vapor as the major absorber of infrared radiation emitted by the earth.

It should be apparent that the higher the concentration of water vapor in the atmosphere, the more earth radiation is absorbed and the higher atmospheric temperatures remain. The absorption of radiant heat by water vapor gives rise to a noticeable difference between the daily temperature range of an arid or semiarid climate and that of a humid climate. During the summer in Colorado, for example, where the average relative humidity may be 35 per cent, a daily temperature range of 30 Fahrenheit degrees is not uncommon. Pleasant nighttime temperatures may follow even very high afternoon temperatures. At comparable latitudes in very humid parts of the country, where the relative humidity might be as high as 90 per cent, the daily range may be only 10 to 15 Fahrenheit degrees. Consequently, there is little difference between daytime and nighttime temperatures, and the result is uncomfortable, warm nights. Atmospheric humidity is the primary factor operating to produce these different conditions.

It is possible to use the absorption of infrared radiation as an indicator of atmospheric humidity. Although the presence of carbon dioxide and ozone may affect the absorption of infrared quantitatively, the percentages of these two gases are relatively unchanging in various air masses, being about 0.03 per cent and less than 0.01 per cent by volume, respectively. On the other hand, the percentage of water vapor varies considerably (from 0 to 7 per cent) with the passage of dry and moist air masses. Thus, variations in water vapor, and consequently in relative humidity, can be detected.

An instrument for studying the transmission of infrared radiation through various materials and gases, including water vapor, is described in the following pages. You will need 2 infrared sources, hot objects that radiate infrared rays. These may be in the form of containers, each with a capacity of about 1 quart, which can be filled with hot water at about 180° F. One container should be made of material that is a good radiator of infrared, such as glass or metal, painted black on the outside. The second container might be of a polished metal, such as stainless steel or chromium-plated steel. A flashlight or spotlight can be a reference source. The range of wavelengths from the flashlight compares with that of sunlight.

For varied test materials, sheets—50 centimeters on a side or larger—of some of the following are suggested: polished aluminum or aluminum foil, glass, plexiglass, polyethylene, heavy fabric, sheer nylon, or silk fabric. These sheets can be placed between the heat source and a detector to note their absorptive effects.

Commercial detectors for sensing radiation are available, but you can make your own sensor in a thermopile, or series of thermocouples. The thermopile should be as small as possible and covered with a thin coat of dull black paint. When radiation strikes the sensor, it is absorbed, and there is an increase in the temperature of the thermopile. It is this rise in temperature that is detected and measured.

A thermocouple consists of 2 electrical conductors of dissimilar metals, joined at the point where heat is to be applied. The free ends are connected to an electrical current-measuring instrument, such as a microammeter or a sensitive laboratory galvanometer. (These can sometimes be found in Army surplus stores.) When heat is applied to the junction of the dissimilar metals, a small electric current is produced and registered by the microammeter. The amount of current produced is proportional to the temperature difference between the warm and cool junctions.

Pairs of thermocouple wires made of copper-constantan or alumel-chromel may be used. These alloys are inexpensive and may be obtained from any electronic supply house. The diagram in Figure 25 shows the arrangement of copper and constantan wires in a thermopile. In order to secure as great a temperature difference as possible, alternate junctions of the pile are exposed to the radiation while the others are protected from it. In the diagram, the wires are shown to be

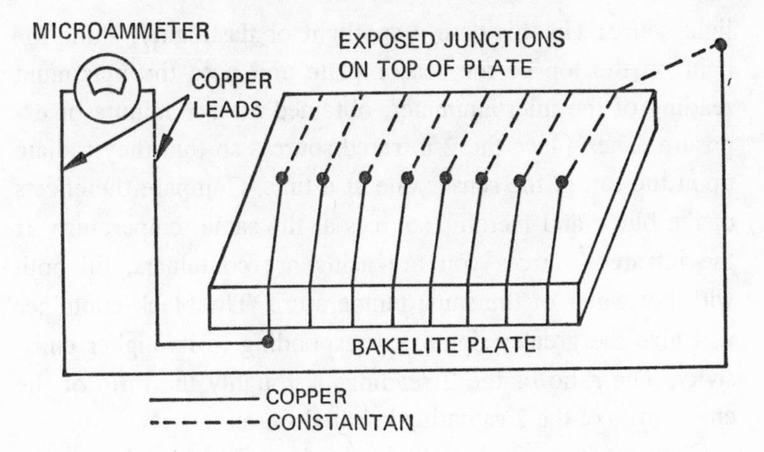

Fig. 25. Thermopile Arranged for Operation

joined around a heavy opaque plate of insulating material, such as bakelite, a material available as scrap from most radio and television repair shops. The junctions exposed to radiation are on top of the plate; the protected junctions are on the bottom and are not visible in the diagram. All junctions should be very small and can be prepared by welding with a propane torch. The entire plate should be sprayed with a dull black paint. Figure 25 shows only 8 pairs of junctions, but 100 or more pairs may be used. The more junctions, the greater will be the sensitivity of the thermopile.

Using the copper wire leads, attach the thermocouple pairs to the microammeter, as shown in Figure 25. The microammeter will register variations proportional to the difference in temperature between the exposed junctions and the protected junctions, and is capable of detecting temperature changes from a few hundredths of a degree to several degrees.

To obtain the intensities of radiation relative to the visible light source (in this case a spotlight or flashlight), shine the light on the top of the sensor plate and note the maximum reading of the microammeter, obtained after 1 minute of exposure. Then place the 2 infrared sources so that they radiate upon the top of the sensor one at a time. Compare the effects of the black and metallic sources at the same temperature. If the infrared sources you are using are containers, fill both with hot water of the same temperature. The black container will give the greater signal, corresponding to its higher emissivity. The ratio of the 2 readings is roughly the ratio of the emissivities of the 2 radiating infrared sources.

Compare the results obtained after exposing the sensor plate to the black infrared source at 2 different temperatures. The ratio of these 2 readings is the ratio of the emissive power of a black infrared source at different temperatures. What is the relationship between temperature and emissivity? For black body radiation, the Stefan-Boltzmann Law states that the intensity of radiation is proportional to the fourth power of the absolute temperature. Stated mathematically, it is:

$I=\sigma T^4$, where

$I=$(emissivity) intensity of radiation in langleys per minute, σ (sigma)=Stefan-Boltzmann constant (8.14×10^{-11} langley per minute per degree Kelvin to the fourth power), and $T=$temperature in degrees Kelvin. (A langley is one calorie per square centimeter, and degrees Kelvin=degrees Centigrade+273.)

To use the Stefan-Boltzmann Law to compare the emissivities of a black infrared source at different temperatures, suppose the readings obtained at the 2 temperatures are 20 microamperes at the lower temperature and 40 microamperes at the higher temperature. Assuming the readings are directly proportional to the temperature difference between the warm and cold junctions, one can set up the following equation:

$$\frac{I_1}{I_2}=\frac{T_1^4}{T_2^4}=\frac{R_1^4}{R_2^4}$$

where the subscripts 1 and 2 refer to the temperature conditions and R refers to the reading of the microammeter. Thus:

$$\frac{I_1}{I_2}=\frac{R_1^4}{R_2^4}=\frac{(2.0\times10^1)^4}{(4.0\times10^1)^4}=\frac{16\times10^4}{256\times10^4}=\frac{1}{16}$$

The emissivities I_1 and I_2 are in the ratio of 1 to 16. From this we can make the general statement that when the readings of the microammeter are in the ratio of 1 to 2, the emissivities will be in the ratio of 1 to 16.

To use the thermopile to study the absorption of infrared radiation by water vapor, obtain relative readings of the microammeter when a polyethylene sheet is placed between the black infrared source and the sensor, and when it is removed. (Polyethylene transmits infrared.) Now dip the sheet in water and obtain a reading when the wet and dripping sheet is suspended between the source and sensor. What is the result?

Further tests may be devised showing the effect of water vapor on the transmission of infrared radiation. These tests will require a method of getting water vapor between the infrared source and the sensor. A good method is to use a 6-inch diameter pile approximately 1 meter in length. Use a stovepipe, if you can find one, or construct a pipe from a sheet of fairly heavy plexiglass and paint the outside dull black. The pipe must have a highly polished interior and polyethylene ends. Water vapor can be introduced through a 1-inch inlet and removed from a similar outlet, as shown in Figure 26. Wet- and dry-bulb thermometers should be mounted at the midpoint of the tube in order to secure readings from

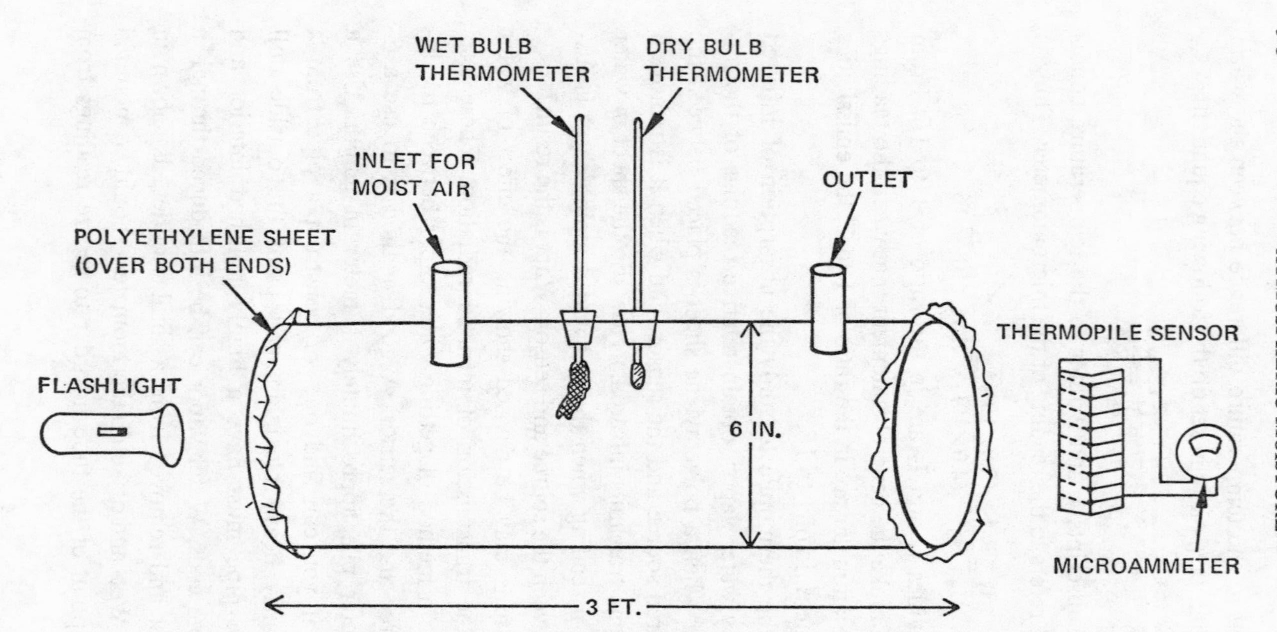

Fig. 26. Apparatus for Studying Effect of Water Vapor on Transmission of Infrared Radiation

which relative humidity can be obtained. To mount the thermometers, drill or cut 2 1-inch holes side by side in the middle of the pipe. Insert 1-hole rubber stoppers, size No. 6, through which identical Fahrenheit laboratory thermometers have been pushed (see Figure 26). To vary the relative humidity, use the apparatus in conjunction with the device for studying critical relative humidities for growth of water droplets on condensation nuclei shown in Figure 28.

Place the infrared source at one end of the pipe and the sensor at the other. Gather data on relative humidities and microammeter readings obtained from the infrared sensor. Can you find a relationship between these 2 sets of data? Try varying the temperature of the infrared source, the length of the tube, and the temperature of the air in the tube.

How can the information you have obtained be used to measure relative humidity in the open atmosphere? Try your apparatus in the open air using temperatures and distances comparable to those you used in the experimental apparatus.

INVESTIGATION 12. (E)

Studying Local Precipitation

MATERIALS NEEDED:

Large-scale map of the area to be studied
About 16 empty cans, 4–6 inches in diameter

Official records of precipitation are obtained by mounting a gauge in an open area away from trees, buildings, or other obstructions. Accumulated records indicate the annual amount of local precipitation for the locality and eventually become part of the climatological summary for the whole region.

Suppose that two precipitation gauges located a mile apart are compared. Would they record the same amount of precipitation for a particular storm? For a month? For a year? Or would the records for the two areas be quite different?

After a summer thunderstorm or shower, you have probably heard people comparing notes: "It rained nearly an inch at our place west of town. How much did you get?"

"The storms always seem to go north of us. Jonesville really got a soaker again today. We didn't get a drop."

What might account for these variations? Do they average

out over a year or is there a preferential pattern of rainfall in some localities?

To study the incidence of precipitation, obtain a large-scale map of your area. Preferably, this should be a detailed map such as the kind provided by the United States Geological Survey. Pick a small area, about 9 square miles, for study and lay out a grid pattern of 3 adjacent rows each made up of 3 boxes, numbering the intersections from 1 to 16. Try to establish the pattern along roads so that each intersection of the grid can be reached conveniently on a regular basis for observation. Or work in conjunction with one or more companions and divide the observation points among you.

Set up as gauges a number of metal cans or other watertight containers, about 4 to 6 inches in diameter. You may be able to get a supply of used cans from a local restaurant. Ground-coffee containers will be adequate. Make sure the cans are clean, smooth down any ragged edges, and place a can at each of the intersections marked on the grid. If your grid covers an area of 9 square miles, you will need 16 such cans. Have all cans well exposed and mounted on stands or boxes above any weeds or other plant growth. Avoid the possibility of having rainwater drip off leaves or deflect off other obstructions directly into the cans.

Pour about 10 drops of liquid glycerine into the bottom of each can after each measurement to minimize the problem of evaporation. Examine all cans on a daily visitation schedule for a period of a month, and record the amounts of precipitation that have occurred. A wooden ruler marked in sixteenths of an inch will serve as a suitable measuring stick. Record all observed showers or thunderstorms, noting the times at which they begin and end. Make a note of the local variations (such as proximity to trees, buildings, lakes, hillsides) at each of the observation sites.

Following a shower or thunderstorm, collect your data and plot the values for the intersections where observations were made, on a blank map of the area. Draw isohyets (lines connecting points of equal precipitation) for suitable intervals, such as 1/16 inch. What does your chart tell you about the local incidence of precipitation during a single storm?

After the monthly data have been collected, plot a similar isohyetal map for the entire month. To do this, total the amounts of precipitation for the whole month at each of the observation points. Can you discover areas of maximum and minimum precipitation in your locality?

After completing the study, try to answer the following questions:

1. How representative is the precipitation value obtained at *one* station over a period of a month? To answer this, determine the average precipitation of all stations on the grid and find the deviations above and below this average for each station.
2. How does summer rainfall vary over a local area of 9 square miles?
3. How do your precipitation totals compare with the official rainfall records of the Weather Bureau for your locality for a given month? What may account for the differences?
4. Do long-range averages show better or poorer correlation with official records?
5. What is the average size of a local shower or thunderstorm? Is a 9-square-mile area large enough for this kind of study? How would you modify the experiment if you were to repeat it?

INVESTIGATION 13. (M)

Photographing Snow Crystals

MATERIALS NEEDED:

½-inch thick plywood, 1 foot by 10 feet
2 small hinges
Black paint
Photopaper holder
Microscope slides
Porcelain or plastic socket
Flashlight bulb, 1½ volts
Push-button switch
1½-volt dry cell
No. 24 plastic-coated copper wire
Photographic paper

Snow, particularly a single snow crystal, is without question one of the most beautiful and unique phenomena occurring in nature. Snow always appears as 6-sided crystals, but no 2 snow crystals are exactly alike. Some crystals are flat; others form long needles. Sometimes the 3 major faces are more prominent than the 3 intermediate faces, giving the crystal a triangular appearance. Clusters of snow crystals can form flakes, pellets, or soft hail.

Many people have never seen the beauty of snow, because it falls on only about one-third of the earth's surface. In some areas, snowfall is so heavy that it may be more of a nuisance than a thing of beauty. But subjected to photographic analysis, the minute realm of the snow crystal can indeed be a "wonderland" full of exciting discoveries.

Obviously, the main problem of any detailed examination of a snow crystal is how to get a good look at the crystal before it disappears forever. An ingenious method for obtaining a permanent, photographic record of snow crystals has been devised by Keiji Higuchi. The device he constructed consists of a light-tight box painted black on the inside with flat black paint. The box has a 1-foot-square base and 2-foot-high sides, and it can be constructed of ½ inch plywood. The bottom of the box is hinged to give access to the interior where the photographic paper will be placed.

A flat holder for photographic paper can be obtained from a photo shop. Attach two small strips of wood with screws to the inside of the hinged base to accommodate the holder, and secure it in position. Insert the paper when you are ready to begin photographing.

Attach a small porcelain or plastic socket for a flashlight bulb to the inside of the top face of the box. Drill two small holes through the top of the box for electrical connecting wires.

To the outside of the top face of the box, attach a small push-button electrical switch and a 1½-volt dry cell (e.g., similar to a Burgess Little Six cell). Clamp the dry cell to the outside of the box with screws and a small metal strip. Use No. 24 plastic-coated copper wire to complete the electrical connections (in series from dry cell to switch to light socket and back to dry cell). Put a 1½-volt flashlight bulb in the socket, and your miniature photographic studio is ready for use (Figure 27).

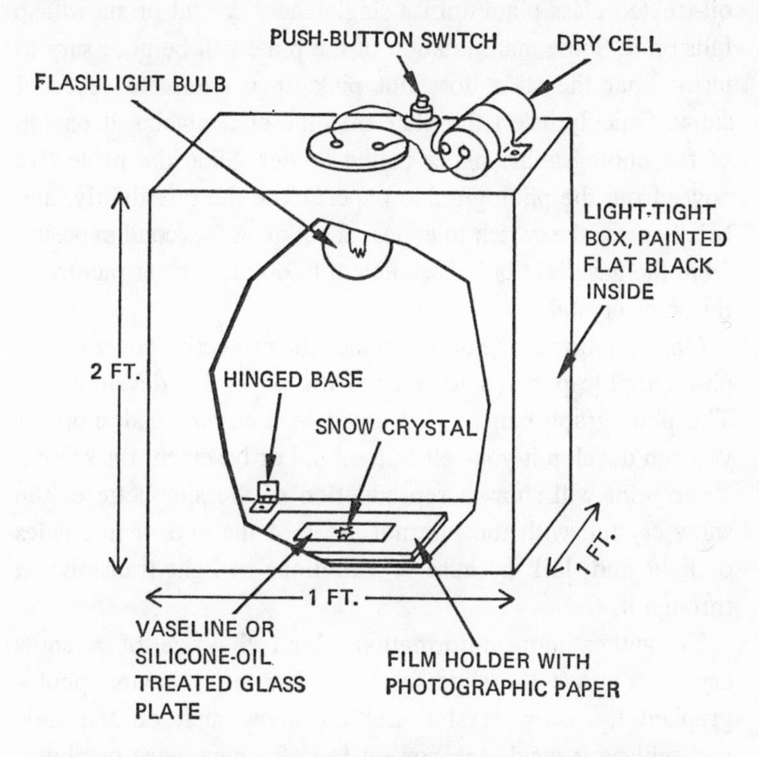

Fig. 27. Photographic Box for Taking Pictures of Snow Crystals

Prepare several standard microscope slides (3 centimeters by 10 centimeters) by coating them with Vaseline or silicone oil. These will be placed on top of the photographic paper one at a time when samples of snow crystals are obtained on them for study.

Set the apparatus outdoors in a protected shelter—an open

porch, for example—so that it can reach the outdoor temperature before exposure of the glass plate to the falling snow. On a dark, snowy night, working by a minimum amount of light—just enough to see what you're doing—expose the silicone oil-treated glass plate until a single snow crystal or snowflake falls on it. Some manipulation of the plate will be necessary to insure that the slide does not pick up a complete cover of snow. Quickly insert the plate into the box, placing it on top of the photographic paper in the holder. Slide the protective cover from the photographic paper, close the box tightly, and briefly press the switch to expose the film. A 3-second exposure from the flashlight bulb is sufficient to obtain a clear picture of the snow crystal.

Open the box in the dark, replace the protective cover on the photographic paper, and remove the paper for development. The photograph can be developed in a commercial shop, or you can develop it yourself in a school or basement darkroom. Your print will show a reproduction of the silhouette of the snow crystal, with the internal areas of the crystal in shades of light and dark because of variations in light transmission through it.

To gather more information about the size of a snow crystal, remove the slide from the box after having photographed the snow crystal. Let the snow melt on the slide and replace it inside the box on top of a new piece of photographic paper. Using the same procedure as before, photograph the melted snow crystal and develop the paper. Measure the diameter of the droplet on the developed photograph with a millimeter ruler and calculate the volume of water in the snow crystal. (Assume that the droplet is spherical and use the formula for obtaining the volume of a sphere: $V=4/3\ \pi r^3$.)

After you obtain several photographs of snow crystals, before and after they have melted, classify them according to size, shape, water volume, or any other criteria you can think of. What variety of forms do you find? Do successive snowstorms show the same variety of snow crystals? How is crystal shape related to temperature at the time of the snowfall? (Keep a record of the outdoor temperature while making your photographs.) What shapes or varieties are produced during blizzards, wet snows, or low temperatures? How do snow crystals compare with other ice forms such as sleet and snow pellets?

INVESTIGATION 14. **(M)**

Observing the Growth of Water on Condensation Nuclei

MATERIALS NEEDED:

1 Erlenmeyer flask
1 Bunsen burner
2 2-hole rubber stoppers, No. 6
2 1-hole rubber stoppers, No. 6
1 small glass vial, 1 inch diameter by 3 inches high
1 small air blower (e.g., a hair dryer)
4 cardboard connecting tubes, 1½ inch diameter
Cardboard box, 1 inch by 1 inch by 10 inches
3 cardboard baffle plates
Styrofoam box, 1 inch by 1 inch by 1 inch
Dry-bulb thermometer, Fahrenheit, −10° F to 110° F
Wet-bulb thermometer, Fahrenheit, −10° F to 110° F
Small cotton wick for wet bulb
1 microscope, medium power
Several glass microscope slides

The importance of condensation nuclei in the formation of rain and snow is undisputed. Much remains to be learned, however, about the exact mechanisms at work and the variations in effectiveness of different types of small particles in the atmosphere. The difficulties of study are compounded by the extremely small size and immense variety of the particles. Also, certain types of nuclei are hygroscopic—they have an affinity for water vapor in the air. With enough hygroscopic nuclei present, condensation of water vapor can occur at very low relative humidities. For example, condensation on sodium chloride nuclei may occur at a relative humidity of 75 per cent. With a more hygroscopic substance, such as lithium chloride nuclei, a relative humidity of only 11 per cent is sufficient to permit condensation. Imagine the vast difference there would be in the total amount of sky cover and precipitation throughout the world if appreciable quantities of lithium chloride were present in the air.

To facilitate the study of different types of condensation nuclei and their relative degree of hygroscopicity, an ingenious device for producing air samples of controlled relative humidity has been devised by James Juisto and Roland Pilie. The entire apparatus consists of 5 parts: a steam generator, water trap, blower, mixing chamber, and relative humidity test chamber (see Figure 28).

The steam generator is simply a flask of boiling water, equipped with inlet and outlet tubes. The first tube will be used to regulate input of the surrounding air and thus vary the level of humidity of the test air. The outlet tube will run to a small water trap, the function of which is to collect excess water and aid in the flow of moist air.

Use a 500-milliliter Pyrex Erlenmeyer flask with a 2-hole rubber stopper, and mount the flask on a support stand about

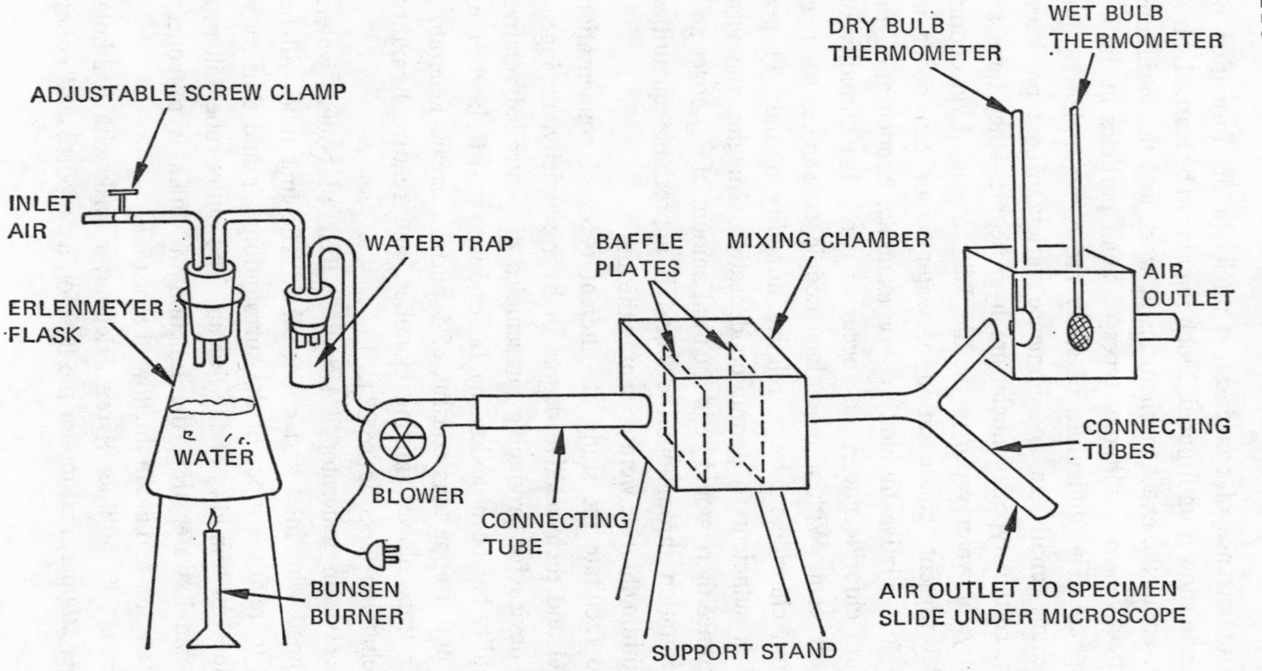

Fig. 28. Variable Humidity Apparatus

7 inches high so that a Bunsen burner can be placed underneath. When you are ready to begin the investigation, the flask should be filled about three-fourths full with water kept at a low boil. Insert 2 bent glass tubes in the rubber stopper. To the inlet tube, attach a short length of rubber tube and a screw clamp.

The water trap is a small flask, about 100 milliliters, with, again, a 2-hole rubber stopper. Glass inlet and outlet tubes protrude from the stopper, the inlet tube connected to the outlet tube of the generator by a piece of rubber tubing.

The outlet tube from the water trap leads to the intake side of a small air blower, which can be a hand-held hair dryer. Many modern dryers have controls for both warm and cool air. Run the dryer on cool, or, if possible, disconnect the electrical heating element from the dryer. Connect the water-trap outlet tube to the dryer with a length of rubber tubing, sealing the joint tightly with masking tape. Position the dryer on a suitable support.

Make a 1-foot-square mixing chamber from cardboard or styrofoam, taping all sides when you are completed to make the box airtight. To force the air through a circuitous path for better mixing, insert 2 or 3 pieces of cardboard, taped or wedged into vertical position, to act as baffle plates.

The relative humidity test box can be approximately the same size and of the same material as the mixing chamber, but the baffles are omitted. Suspend wet- and dry-bulb thermometers (see Investigation 4), mounted side by side, loosely through rubber stoppers in the top of the box for easy removal. This chamber will be the control, enabling you to gauge the relative humidity of the air that reaches the condensation nuclei specimen. Both the relative humidity test box and the mixing chamber should be mounted on support stands.

The various connecting tubes may be fashioned from cardboard mailing tubes or 1-inch diameter glass tubes. The outlet from the relative humidity test chamber can lead directly to the stage of a high-power optical microscope through which you will observe the condensation process; or a branching tube from the mixing chamber, as shown in Figure 28, can serve this purpose. The microscope will be prepared with a slide containing the condensation nuclei to be examined.

Using this device, it is possible to produce output air with relative humidity ranging from a minimum equal to that of the ambient air in the room, to a maximum of 98 per cent. The degrees of relative humidity are achieved by regulating the input of room air with the adjustable pinch clamp at the inlet. Slight increases in the amount of room air to the water vapor will change the relative humidity by small increments throughout the entire range.

To collect condensation nuclei for observation, expose clean, glass microscope slides to the air for a period of 1 to 2 hours during quiet conditions. The nuclei collected in this manner will be representative of the particles in suspension in the atmosphere.

(Another method for collecting nuclei is to attach narrow glass slides to the rotating head of an electric hand drill by using a small metal T-frame in which 2 slides are mounted at the ends of the crossbar. Rotate this rapidly outdoors on a foggy day. After the droplets or crystals have evaporated, a large number of condensation nuclei will remain on the slide.)

A study of condensation nuclei under these controlled humidity conditions will reveal that as a certain level of humidity is approached, the edges of the hygroscopic particles begin to dissolve and become indistinct. Eventually, a point is reached at which the hygroscopic nuclei dissolve in the absorbed moisture. The relative humidity observed at this point is called the critical humidity and is an indication of the degree of hygroscopicity of the particles under observation. Various condensation nuclei have different critical humidities and, as mentioned previously, more hygroscopic particles dissolve under lower relative humidity conditions. Most particles will dissolve between 75 per cent and 100 per cent relative humidity. Some nuclei, however, are non-hygroscopic and will remain unchanged throughout all ranges of relative humidity.

Once you are satisfied that the apparatus is functioning properly and have examined specimens of nuclei in normal air, run a series of comparison studies. Sodium chloride crystals can be prepared by crushing table salt between 2 microscope slides. Seventy-two to 75 per cent relative humidity is the level at which condensation usually begins with sodium chloride nuclei. Collect nuclei samples from seawater, if possible, by letting ocean spray strike and evaporate on a microscope slide. Protect the slide with a foil covering until you reach your testing equipment and determine the critical humidity. Does this study suggest any ideas about moisture in the atmosphere near beaches?

Examine minute particles of dust, pollen, or soil for their potential as condensation nuclei.

Conduct a series of daily investigations, taking into consideration the time of day, season, prevailing weather conditions, visibility, and other factors. Can the extent of air pollution be related to these studies?

Chapter 4

ATMOSPHERIC MOTIONS AND CIRCULATION

In the story *Around the World in Eighty Days,* a balloon was the principal means of transportation. Suppose one could ride the gondola of a balloon for 80 days. Could he circumnavigate the globe in that period of time? How long does it take for a given mass of air to completely circle the earth? In which direction does it travel?

A joint project by the National Center for Atmosphere Research (NCAR), the Environmental Science Services Administration (ESSA), and the New Zealand Weather Service is investigating similar questions. The project is called GHOST, for Global Horizontal Sounding Technique, and involves sending plastic constant-density balloons on extended flights around the globe. These balloons carry telemetry equipment capable of sending electronic signals to satellites and ground bases. Most of the flights have been made in the Southern Hemisphere, some lasting as long as a year and circling the earth dozens of times. When meteorologists are able to send aloft about 10,000 of these balloons, keep them up for periods up to 2 years, and track them continuously for that period, the project will be considered truly operational and will add immensely to the storehouse of information about circulation of the atmosphere. But much has already been learned about the major atmospheric movements that lie behind the prevailing wind patterns in any particular locality.

We can think of the atmosphere as an enormous engine using a moving fluid, air, to do work. The energy for this engine comes from the sun through radiation that is trapped by the atmosphere and changed into heat. Since the source of this heat is more abundant in equatorial regions than in polar regions because of the directness of the sun's rays, the atmospheric "engine" operates to carry heat poleward in each hemisphere. Gravity acts as the basic driving force, producing convection currents as the average air density is decreased in equatorial regions due to heating. The heated air, pushed upward by cooler air, spreads northward and southward, and eventually settles toward the earth's surface again at higher latitudes. In time, this air returns to the equatorial regions to complete the cycle (see Figure 29).

The general circulation pattern of the atmosphere, prevailing wind belts, and semipermanent pressure cells result from this balance between the excessive incoming radiation near the equator and excessive outgoing radiation in polar regions. On the average, the dividing line between these two excesses is at approximately 35° to 40° latitude in each hemisphere. In the lower latitudes, more radiation is received than returned, while in higher latitudes the reverse is true.

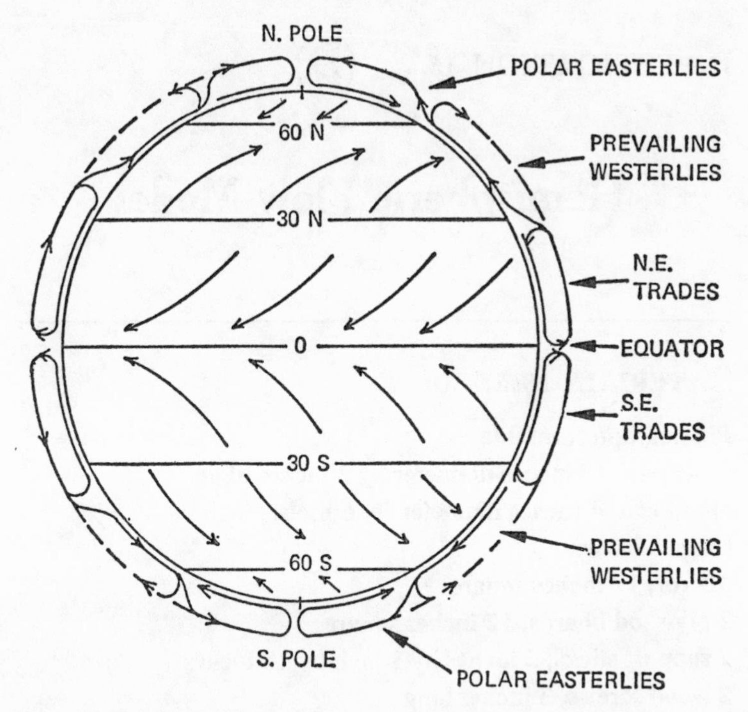

Fig. 29. General Circulation Pattern and Prevailing Wind Belts of the Earth

But several other phenomena exert varying effects on air movement: the rotation of the earth, topographic features (such as mountains and water masses), and the presence of water vapor in the atmosphere. The effect of rotation on circulation patterns is the subject of the following investigation.

INVESTIGATION 15. (D)

Hemispheric Flow Model

MATERIALS NEEDED:

Phonograph turntable
Cake pan, 10 inches diameter by 2 inches high
Small can, 4 inches diameter by 6 inches high
Bunsen burner
Mirror, 12 inches square
2 plywood boards, 12 inches square
2 support sticks, 2 inches by 1 inch by 18 inches
2 wood screws, 3 inches long

For many years, meteorologists have used models to simulate the large-scale motions of the earth and its atmosphere. A geographical globe is, of course, a model in miniature, suitable for studying solid portions of the earth, such as the relative positions of continents and seas, and the earth's rotational characteristics. But, to study the atmosphere, something flexible is needed, to which various experimental conditions can be applied in an effort to simulate what happens in the real atmosphere.

What is the relationship between the rotational speed of

the earth and the number of high- and low-pressure areas, or cells, present at a given latitude? What effect does the temperature gradient between equator and poles have on the frequency of cell development? What is the effect of the earth's rotation on the development of weather patterns in the Northern Hemisphere? These are some questions that may be studied more effectively through a suitable model.

In 1957, while conducting experiments at the University of Chicago, Dr. David Fultz initiated the use of the "dishpan" model to study the atmospheric wave patterns within the troposphere, the lowest 30,000 to 50,000 feet of the atmosphere. The "dishpan" model simulates a hemispherical section of the atmosphere as a flattened earth contained in a pan or pot. The pan is filled with liquid, representing the atmosphere, and turned very slowly on a phonograph turntable. A container of ice placed in the center of the liquid simulates the cooling effect of the earth's poles, and a flame applied to the rim of the pan will serve as the sun's heat. Due to the Coriolis effect, the turntable motion deflects any particles floating in the liquid, and the observer gets a fairly representative picture of the broad-scale atmospheric movements of the earth. This deflection is named for French mathematician G. G. Coriolis, who investigated it in 1835; it is caused by the combined rotational motion of the earth and the movement of the air relative to the surface of the earth.

Coriolis acceleration results when a rotary (circular) motion and a translational (straight) motion act simultaneously upon a moving object. The combined effect of these 2 motions makes it appear as if an object moving in a straight line is being subjected to a sideways force, causing it to change its direction. In reality, the 2 motions are independent, but the observer sees them as one motion along a curved path.

On the rotating earth, the Coriolis effect helps to create large-scale prevailing wind belts, clockwise (anticyclonic in the Southern Hemisphere) and counterclockwise (cyclonic in the Northern Hemisphere) movements of air masses.

Several factors must be carefully controlled in this model to produce a pattern resembling atmospheric waves. The depth of the fluid in the pan, the type of heat source at the rim, the size of the ice container at the center, and the rotational speed of the turntable are all crucial factors.

Figure 30 illustrates how the apparatus should be arranged. First, find a 4-inch-diameter can—about the size of a large juice can—and a pan 10 inches in diameter and about 4 inches deep. You may be able to locate a cake pan with these dimensions, which is an ideal container for the model. A 4-quart metal casserole or baking pan, round and fairly straight-sided, should fit the size requirements adequately. Place the pan in the center of the turntable, which will have to be one with a removable spindle (or 2 washers can be placed around the spindle and the can placed on top of them).

Fill the can with crushed ice and place the can carefully in the center of the pan. Exact centering is important here to avoid sloshing, which will destroy the effects you are looking for. Fill the pan with water to a depth of 1½ inches and begin rotating the turntable at about 7 rpm (revolutions per minute).

To achieve this slow speed of rotation (between 5 to 10 rpm), you will have to devise a method of rotating the turntable about one-fourth as fast as 33⅓ rpm, generally the slowest speed on most modern record players. It is possible to modify the phonograph's drive mechanism to reduce the speed of rotation. But a homemade "stopper" device will

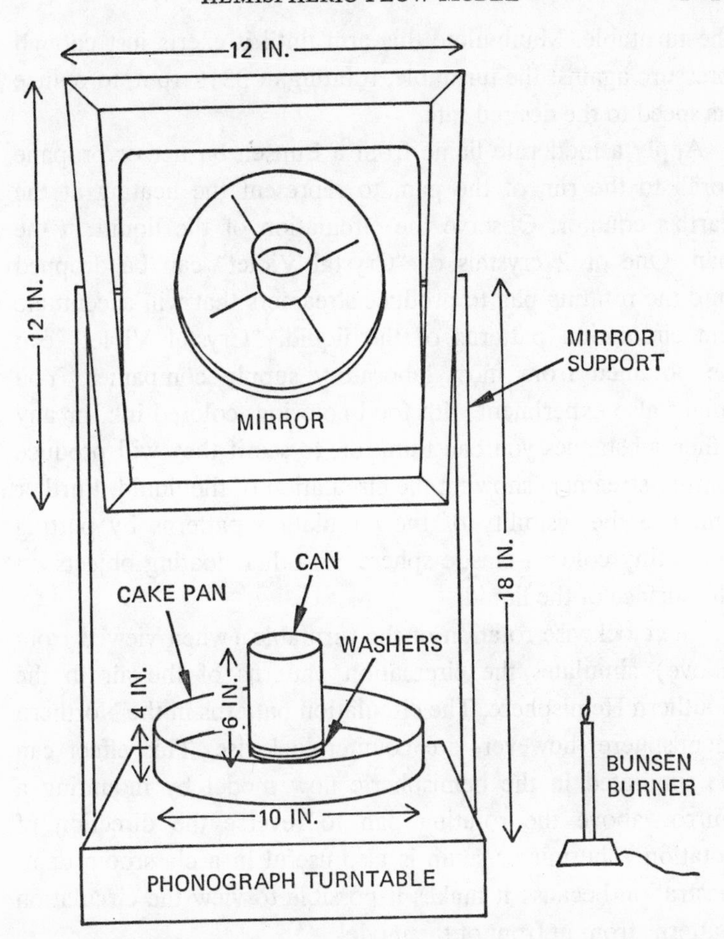

Fig. 30. Hemispheric Flow Model

achieve the same result. Position the turntable near enough to the edge of the table to accommodate a "C" clamp on one side. Insert between the clamp and the turntable side a piece of sturdy rubber, such as that used in cooking spatulas, or other material, positioning the rubber so that it just touches the turntable. Manipulate this arm until it exerts just enough pressure against the turntable, rotating at 33⅓ rpm, to reduce its speed to the desired rate.

Apply a moderate flame from a Bunsen burner or propane torch to the rim of the pan, to represent the heating at the earth's equator. Observe the circulation of the liquid in the pan. One or 2 crystals of "Crystal Violet" can be dropped into the rotating pan to produce streamers that will accentuate the circulation patterns of the liquid. "Crystal Violet" can be obtained from most laboratory supply companies. You might also experiment with food coloring, colored ink, or any other substances you can think of, to see if they will produce similar streamers showing the circulation of the liquid. Further enhance the visibility of the circulation patterns by putting some tiny colored-plastic spheres or other floating objects on the surface of the liquid.

The clockwise rotation of the turntable (when viewed from above) simulates the circulation patterns of the air in the Southern Hemisphere. The circulation patterns in the Northern Hemisphere, however, are counterclockwise. This effect can be simulated in the hemispheric flow model by mounting a mirror above the rotating pan to reverse the direction of rotation. The mirror setup is also useful in a classroom demonstration because it makes it possible to view the circulation patterns from in front of the model.

After the pan is set into rotation, 4 separate stages in cell development are discernible, as illustrated in Figure 31. When the pan is first set in motion, a period of time elapses before the water is accelerated to the speed of the pan itself. During this first stage, any dye streams in the liquid will indicate a flow in concentric circles about the pole—the ice-filled can (see Figure 31a).

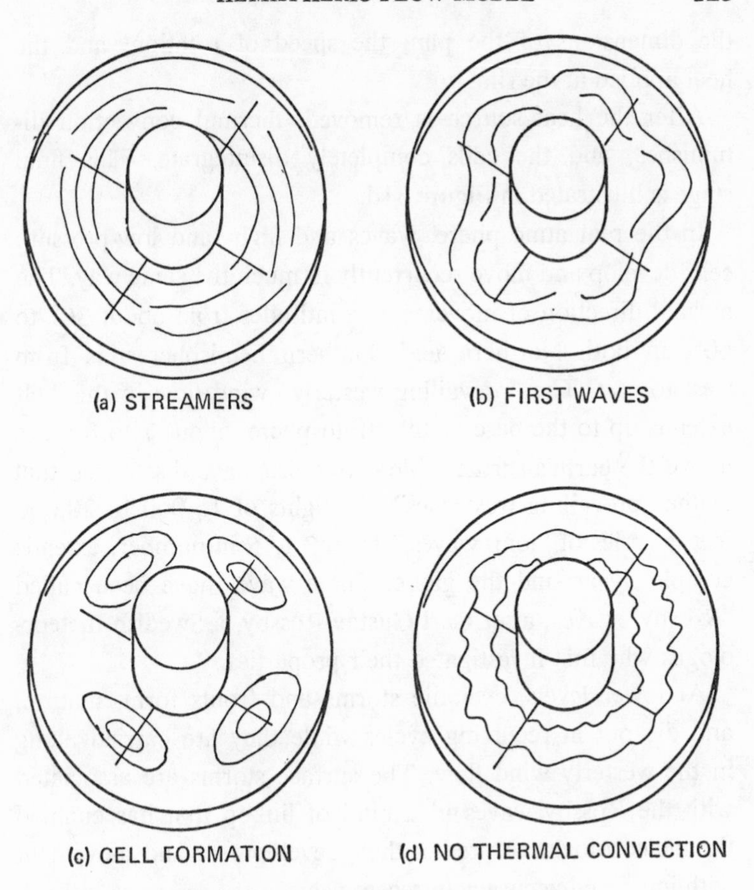

Fig. 31. Hemispheric Flow Development

As the effects of heating at the rim (equator) and cooling at the center (pole) became apparent, there is a transition from zonal flow in concentric circles to long, continuous waves, as shown in Figure 31b. These waves may become further defined as separate cells (Figure 31c), although the wave pattern is more common. The number of waves or cells that will form, as well as their amplitude, depends upon the dimensions of the pan, the speed of rotation, and the heat applied at the rim.

After the heat source is removed, thermal convection diminishes, and the cells completely disintegrate. This final stage is illustrated in Figure 31d.

In the real atmosphere, waves and high- and low-pressure cells develop and move recurrently in much the same way. The normal direction of movement in latitudes from about 30° to 60°, in both Northern and Southern hemispheres, is from west to east. The "prevailing westerly" wind flow in this belt extends up to the base of the stratosphere, about 7 to 8 miles above the earth's surface. Meteorologists have discovered that in the "prevailing westerlies" at heights of 15,000 to 20,000 feet, a series of "long waves," from 3 to 5 in number, extends completely around the globe. These waves have been called "Rossby" waves, after Carl Gustav Rossby, a Swedish meteorologist who first investigated their properties.

At lower levels, cyclonic storms and fronts form, mature, and die out in recurring cycles while they are carried along in the westerly wind flow. The surface storms are associated with the Rossby waves in a kind of linkup that has enabled the meteorologist to predict their development and movement with improved accuracy in recent years.

At the ground, we reap the varied forms of weather conditions associated with the fronts and cyclones. The word "cyclone" does not necessarily mean a violent storm, but refers to a low-pressure center with counterclockwise movement of air in the Northern Hemisphere. (The direction of rotation is reversed in the Southern Hemisphere.) Frequently, rain or snow, thunderstorms, gusty winds, and overcast skies are associated with cyclonic storms. The boundaries between significantly different air masses (different in temperature, moisture content, and other factors) in the cyclone are called "fronts," and it is along these boundaries that the real "battle of the elements" takes place.

The "dishpan" model of the atmosphere is an attempt to study the real atmosphere on a miniature scale. But since the fluid used in the "dishpan" model is not air, but water, many differences exist between the model and the atmosphere. To use this model for any kind of meaningful study requires translating the results obtained into terms that apply to the atmosphere. In other words, the results must be "scaled up" to fit the atmosphere.

Several factors of comparison between the model and the atmosphere must be considered: (a) the driving force (the temperature gradient between the hot and cold sources), (b) the deflection due to rotation, (c) the length-to-height ratios of the model and the atmosphere, and (d) the comparative viscosity of the liquid and air.

In working out the scaling factor for this model, Bellaire and Stohrer (see Bibliography) have noted the significant points of divergence. Three ratios of measurements were considered. The first of these is the ratio of the temperature gradient (driving force), $\Delta T/\Delta r$. to the rate of rotation (deflecting force, represented by omega, Ω). The temperature gradient is expressed as the difference in temperature, ΔT, between the hot and cold sources, whose distance apart is Δr. This ratio can be stated mathematically as $\frac{\Delta T}{\Delta r} + \frac{1}{\Omega}$.

The second consideration is the ratio of the radius to the height of the fluid. In the atmosphere, the radius is the distance from the equator to the pole, and the height is the representative height of the tropopause. The tropopause is the top of the troposphere or active atmospheric layer nearest the ground.

The third consideration is the reciprocal of the relative viscosity of the fluid, $1/\mu$ (one over mu). Consider this to be 1 for the atmosphere.

The full mathematical statement of all these factors is:

$$\frac{\Delta T}{\Delta r} \times \frac{r}{h} \times \frac{1}{\Omega} \times \frac{1}{\mu} = \text{scaling parameter}$$

For the atmosphere, the temperature gradient is about 125° F in 6,000 miles (3×10^7 feet), the ratio of radius to height of the tropopause is about 10 miles (5×10^4 feet) in 6,000 miles, the rate of rotation is 360° per day (0.25° per minute), and the relative viscosity is 1. Using this information, the scaling parameter for the atmosphere is:

$$\frac{125^\circ \text{ F}}{3 \times 10^7 \text{ feet}} \times \frac{3 \times 10^7 \text{ feet}}{5 \times 10^4 \text{ feet}} \times \frac{1}{0.25^\circ/\text{minute}} \times \frac{1}{1} = \frac{1}{100}$$

In comparison, using the same temperature difference, the pan measurements and rotation rate, and a relative viscosity of 100 for water, the scaling parameter for the model is:

$$\frac{125^\circ \text{ F}}{0.25 \text{ feet}} \times \frac{0.25 \text{ feet}}{0.125 \text{ feet}} \times \frac{1}{2{,}500^\circ/\text{minute}} \times \frac{1}{100} = \frac{1}{250}$$

From this, the ratio of the parameters of atmosphere to model is 2.5 to 1. In other words, if your model were operated as described above, you could expect your observed results to differ from the real atmosphere by a factor of 2.5 to 1.

What changes could be made in the model to give a ratio of parameters closer to 1 to 1? Suppose you were able to make the necessary changes to secure a ratio of 1 to 1. Would this indicate a perfect relationship between the model and the atmosphere? What are some hidden defects in this reasoning?

Here are some other questions to consider while experimenting with the hemispheric flow model:

1. What is the advantage of having the heat source at the rim and the cold source in the center? Could it be the reverse? In an experiment by Fultz and Kaylor (see Bibliography), the sources of heat and cold were reversed. What might be the possible results?
2. How are "Rossby" waves expressed mathematically? What is their significance in the atmosphere? Are they of value in meteorological forecasting? Any standard textbook in meteorology, such as *Introduction to Meteorology* by Sverre Petterssen, will describe these waves in more detail.
3. Locate the "frontal zones" on photographs taken of the cyclonic and anticyclonic cell formation in the model. Which cells may be analogous to low-pressure cells in the atmosphere? To high-pressure cells? How can you determine this?
4. What happens to the pattern when one or more of the essential ingredients is missing, for example, heat source, rotary motion, or adequate depth of liquid?
5. Substitute some other clear liquid, such as glycerin, for the water. What differences do you observe?
6. What is the maximum number of stable cells you can form with your apparatus? Describe the sequence (or show photographs) of the breakup and re-formation of the cells. Analyze the steps of growth and disintegration. Does this information show any similarity to atmospheric conditions?

INVESTIGATION 16. (D)

Demonstration of Deflection Due to Coriolis Effect

MATERIALS NEEDED:

Large round-bottom pan, 12 inches diameter, 5 inches depth
¼-inch plug, connected to chain
3–4 support blocks or bricks, 8 inches high
Vegetable coloring

Study of Deflection in Hemispheric Flow Model

The Coriolis effect, we have said, leads to the large-scale prevailing wind belts over the earth. Air in motion across the surface of the earth is deflected to the right or left depending upon the hemisphere in which it is moving. Because the earth turns in a counterclockwise direction, as viewed from a point above the North Pole, air deflection is to the right in the Northern Hemisphere and to the left in the Southern Hemisphere.

Consider a stationary mass of air at 30° N. Lat. on the surface of the earth. At that latitude a point on the earth is traveling about 900 miles per hour due to the daily rotation.

The "stationary" mass of air is also traveling at that speed because it is moving with the earth on its daily journey of rotation. We say, however, that the mass of air is stationary because it is not moving relative to the earth's surface.

Now imagine that the mass of air is caused to move northward, perhaps because of the development of a pressure difference between 2 points in the atmosphere. As the air mass moves northward, its momentum continues to carry it eastward at nearly 900 miles per hour (the speed it had due to the earth's rotation at 30° N. Lat.). There will be some loss of momentum due to the effect of friction as the air moves over the surface of the earth. When the air arrives at latitude 45° N., for example, where the earth's rotational speed is only about 735 miles per hour, it will have moved eastward some distance because of having traveled faster than the earth's surface (i.e., 900 miles per hour for the air mass versus 735 miles per hour for the earth's surface). Therefore, the air's path on the surface of the earth is curved toward the east (see Figure 32). We say that the air mass has "veered" toward the right. When a mass of air moves southward in the Northern Hemisphere, it will similarly veer toward the right. As it moves toward a portion of the earth's surface that has greater rotational speed (nearer the equator), it will fail to keep even with the earth's speed and therefore will end up to the west of its original meridian. In the Southern Hemisphere, the result of the two motions—rotation of the earth and translation across the surface of the earth—produces a veering toward the left. As a consequence of this difference in deflection due to the Coriolis effect in the two hemispheres, similar pressure systems rotate in opposite directions. For example, a low-pressure system (cyclone) in the

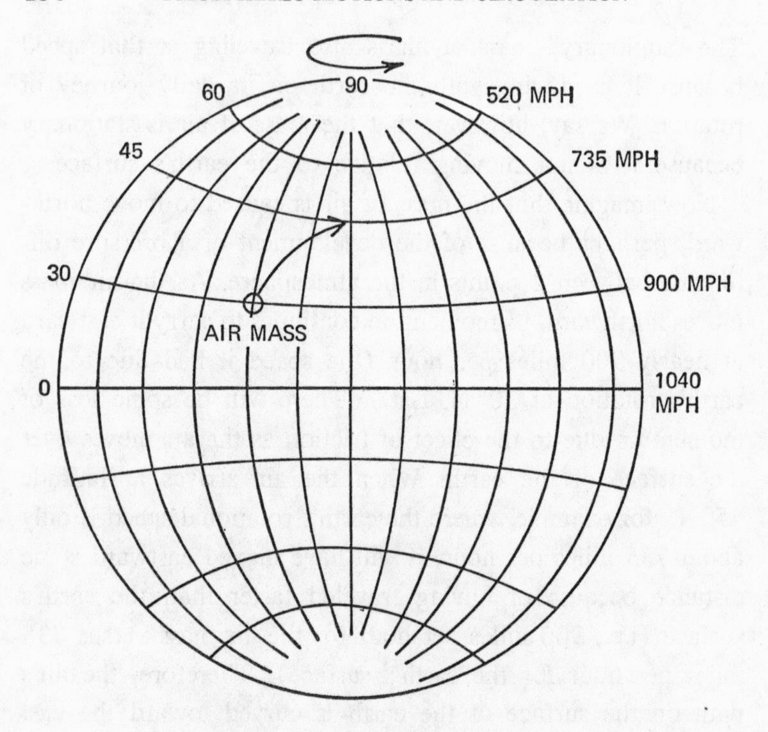

Fig. 32. Deflection Due to Coriolis Effect

Northern Hemisphere rotates counterclockwise, whereas a similar system in the Southern Hemisphere rotates clockwise.

To study the Coriolis effect in greater detail, use the hemispheric flow model described in the previous section. In that experiment, the Coriolis deflection was produced by the rotation of the pan. In the atmosphere, the Coriolis deflection affects all air motion but is minimal in equatorial regions. By using the same apparatus, but with the ice container and Bunsen burner removed and a faster speed of rotation, the Coriolis deflection can be demonstrated with simulated atmospheric motions in any direction.

Rotate the phonograph turntable at a normal speed (i.e., 33⅓ rpm. Set 2 small, narrow juice cans diametrically across from each other inside the pan, each with 2 small holes drilled in one side about 1 inch above the bottom. When the demonstration model is functioning, fill the cans with colored water. As the colored water runs out through the small holes into the "atmosphere" of water in the cake pan, the visible patterns simulate certain atmospheric motions (see Figure 33).

1. With this apparatus, can the Coriolis effect be demonstrated for east–west motion as well as for north–south motion? What about vertical motions in the "atmosphere" (see Figure 33c)?
2. The mathematical expression for Coriolis acceleration for a unit mass of fluid is:

   $a_c = 2\Omega V \sin \phi$, where a_c represents Coriolis acceleration, Ω is the angular velocity, V is the translational velocity of the fluid, and ϕ is the latitude.

   Calculate the Coriolis acceleration for the rotating cake pan with the following measurements:

   $\Omega = 30$ revolutions per minute $= \frac{1}{2}$ revolution per second $= 180°$ per second $= \pi$ radians per second
   $V = 10$ centimeters per second (ejection velocity from the juice cans) Let $\phi = 90°$ ($\sin \phi = 1$)

   Would the Coriolis acceleration change by using a larger pan, say, 20 inches in diameter? What would happen if the rate of rotation were increased to 45 rpm? (See Appendix A, page 240.)

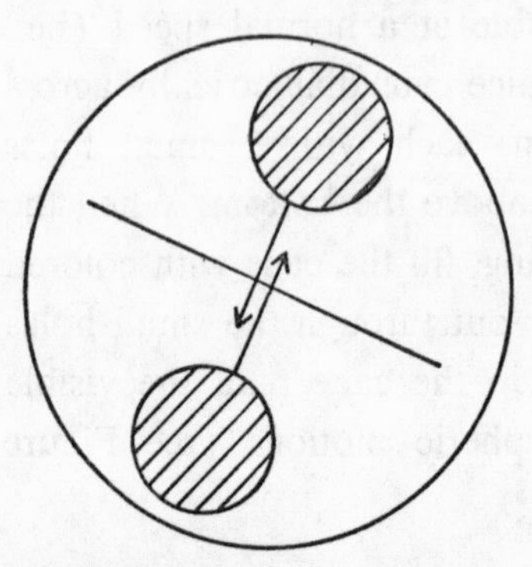

(a) No rotation. Water streams from small cans straight toward each other. Represents absence of Coriolis deflection.

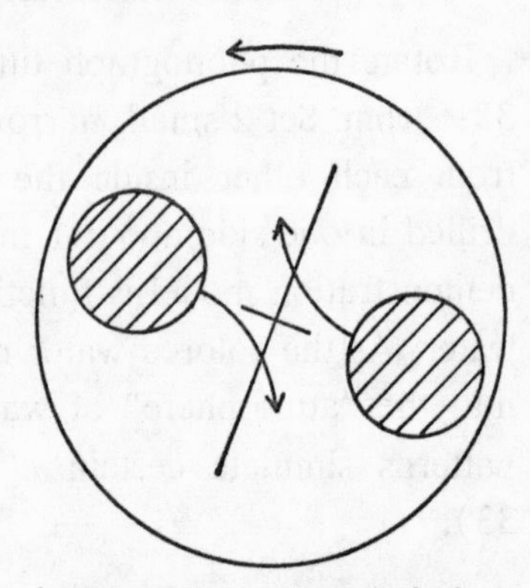

(b) Pan rotates counterclockwise. Streams of water veer toward the right of their initial direction. Represents eastward deflection of a south wind in the northern hemisphere.

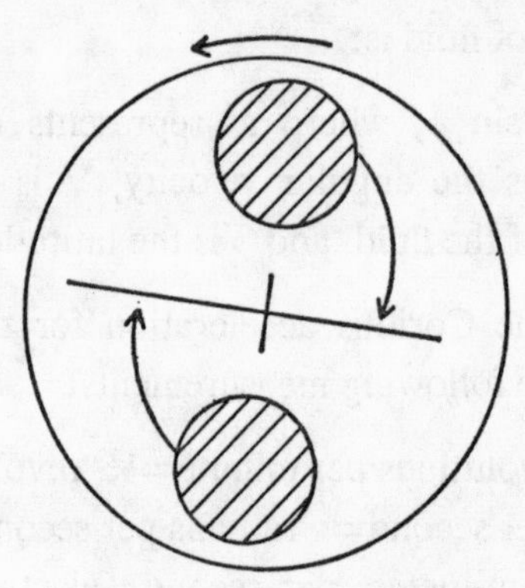

(c) Pan rotates counterclockwise. Streams of water veer toward the right of their initial directions. Represents northward deflection of an east wind in the northern hemisphere.

Fig. 33. Coriolis Demonstration

3. How does Coriolis deflection affect the movement of objects on or above the surface of the earth, other than air, such as projectiles, ocean currents, satellites, etc.?
4. Why are the paths of planes, trains, automobiles, and boats not subject to corrections for Coriolis deflections?

Coriolis Effect on Water Draining from a Sink

Imagine a hypothetical trip across the equator in the cabin of a ship, in which there are no portholes. How could the occupants of the cabin determine when they had crossed the equator? According to the old story, the solution to this problem would lie in draining some water from the sink and watching its direction of rotation as it goes down the drain. Presumably it would rotate counterclockwise in the Northern Hemisphere and immediately reverse its direction of rotation as the ship moved into the Southern Hemisphere. Does the Coriolis deflection affect water draining out of a sink? How could you find out?

For this investigation, you will need a circular and dispensable dishpan with a round bottom, approximately 12 inches in diameter and 5 inches deep. Drill a small hole in the bottom of the pan at the exact center and fashion a removable plug out of cork or rubber, to be inserted from the outside. The pan should be mounted above the drainboard of a sink so that the water from the pan outlet will drain into the sink. A small wooden frame can be constructed to hold the pan securely, or if the pan has a rim it can be supported by bricks or blocks of wood, as shown in Figure 34.

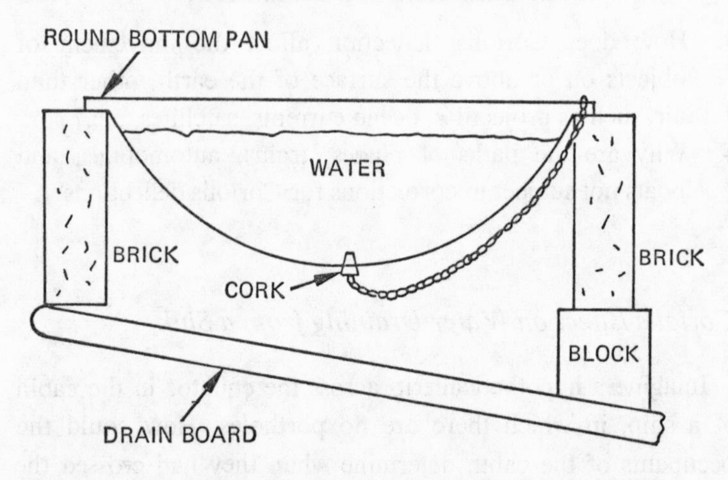

Fig. 34. Device for Studying Water Draining from a Pan

Fill the pan with water to a predetermined level. Allow the water to come to complete rest. (This may require a half hour or so in an undisturbed room.) To be absolutely certain that *all* internal motions of the water have stopped, you can add powdered aluminum or graphite to the surface of the water.

After you are satisfied with the initial conditions of the water, pull out the plug and observe the final direction of rotation of the water as it drains from the hole in the pan. Repeat the operation until you are sure of your results. What is the probability of rotation in each direction? What was the percentage of rotation in each direction after many trials? Is there any built-in bias in the apparatus? How can such a bias be eliminated?

Other questions you might consider are: Why doesn't the Coriolis effect make the water rotate before you pull the plug? (Remember that two kinds of motion are needed to produce the Coriolis effect.) Why does pulling the plug suddenly create an unbalanced force? To what kind of pressure system in the atmosphere is this analogous? (The fluid is flowing inward toward a center.) If your data show approximately equal clockwise and counterclockwise rotations as the water drains out, the question arises as to what causes rotation in the first place. How does the conservation of angular momentum enter the picture? Conservation of angular momentum refers to the tendency of rotating bodies to speed up or slow down as their radius of rotation shortens or lengthens respectively. Recall what happens to a pirouetting skater as she brings her arms closer to her sides. To see what happens when water in the pan has some initial motion, swirl it gently with your fingers just before pulling the plug. What happens to its speed and direction of rotation as it drains out the hole in the pan?

Could you calculate the Coriolis acceleration for a rotating carrousel? Could you show preferential rotation in one direction or another of water draining from a pan while mounted on a rotating carrousel?

To compare the effects of the earth's rotation on a draining pan of water and the effects of a slowly rotating carrousel on a similar pan of water, assume that 2 pans are placed at the earth's North Pole and the center of the carrousel respectively, and that the speed of water flow toward the drain hole is $1/10$ meter per second in each case. If the carrousel is rotating once per minute and the earth rotates once per day, the comparison could be calculated as follows:

Earth	*Carrousel*
$\Omega=2\ \pi$ rad./day (1 rotation per day)	$\Omega=2\ \pi$ rad./min. (1 rotation per minute)
$=0.726\times10^{-4}$ rad./sec.	$=0.104$ rad./sec.
$\phi=90°$ latitude, $\sin\phi=1$	$\phi=90°$ latitude, $\sin\phi=1$
$V=10^{-1}$ m./sec. (1/10 meter per second)	$V=10^{-1}$ m./sec. (1/10 meter per second)
$a_c=2\ \Omega\ V\sin\phi$	$a^1_c=2\ \Omega\ V\sin\phi$
$a_c=(2)(.726\times10^{-4}\ \text{sec.}^{-1})(1)$	$a^1_c=(2)\ (.104\ \text{sec.}^{-1})(10^{-1}\ \text{m. sec.}^{-1})(1)$
$=1.452\times10^{-5}\ \text{m. sec.}^{-2}$	$=.208\times10^{-1}\ \text{m. sec.}^{-2}$

The ratio of the 2 accelerations is:

$$\frac{a^1_c}{a_c}=\frac{.208\times10^{-1}\text{m. sec.}^{-2}}{1.452\times10^{5}\text{m. sec.}^{-2}}=\frac{.14\times10^{4}}{1}=\frac{1{,}400}{1}$$

Therefore, the Coriolis effect of rotation on a pan of draining water for a current of ⅒ meter per second moving toward the drain hole is 1,400 times greater on the carrousel than on the earth. A pan of water draining while on a rotating carrousel should show a preferential direction of rotation due to Coriolis effect, provided that extraneous vibrations and movements can be eliminated. Because the area of the pan is so small and the mass of water relatively large for such a small area, it is unlikely that the earth's Coriolis effect will cause a noticeable effect on water draining from the pan.

We have used models in this section to study the effects of rotation on the atmosphere. Not all models require physical equipment. Reference was made earlier to mental models, which involve abstract concepts and employ mathematical equations to represent certain effects in the atmosphere. It is possible to consider the atmosphere as a single system controlled by physical laws. Mathematical models that take into consideration most of the physical variables that affect the atmosphere, such as energy sources and sinks, topography, and interactions between earth and air, are now being used to predict future states of the atmosphere. The study of such problems has been greatly facilitated by the use of high-speed computers that can solve complicated equations rapidly.

The meteorologist is interested in predicting the weather accurately. Up to now, he has been hampered by slowness in observing current weather, communicating the information, and making the required calculations in time to issue a useful forecast. All these factors have improved gradually over the years, and recent advances in automated observing stations, worldwide communication by satellites, and ultrahigh-speed computers will refine the science of weather forecasting even further.

INVESTIGATION 17. (M)

Model of a Tornado Vortex

MATERIALS NEEDED:

4 posts, 1½ inches by 1½ inches by 16 inches
4 glass panels, 12 inches by 16 inches
2 plywood squares, 15 inches by 15 inches
2 copper tubes, ¼-inch diameter by 6 inches long
1 heater coil, socket, and connecting electrical cord
1 plexiglass flue pipe, 10-inch diameter by 12 inches long; 1 plexiglass flue pipe, 9½ inches diameter by 12 inches long
Wire rack, 10-inch diameter
1 small spotlight, socket, and connecting electrical cord
Pizza tin, 12-inch diameter
Glass or plexiglass circle, 3-inch diameter
Small metal or wooden box, 4 inches by 4 inches by 4 inches
Syringe bulb
Smoke pellets
Dry ice, 1 pound
Aluminum foil, heavy gauge

One of the most awesome spectacles in nature is a tornado. In a well-developed tornado there seems to be almost unlimited power available for destruction. Measurements of the wind velocities, pressure, and temperature changes in the heart of the tornado are often lacking because instruments are destroyed and because of the capricious and transitory nature of the storm. Placement of instruments in suitable positions for recording data is almost impossible because of the relatively small size (perhaps only ¼ mile in diameter) and widespread area over which a tornado may occur.

The causes for this violent and spectacular phenomenon are many and varied. Tornadoes are usually formed in connection with strong cold fronts and squall lines, or with thunderstorms. The central part of the United States is particularly vulnerable, but nearly every state has experienced at least one tornado. Other continents as well, including Africa and Australia, have recorded this type of storm.

Certain conditions favor the formation of a tornado in the United States: (A) very warm, moist air near the surface of the ground, frequently moving northward from the Gulf of Mexico; (b) a cold front moving from the west or northwest toward the warm air mass, causing lifting and marked instability of the air; (c) high-velocity winds at 10,000 to 20,000 feet blowing above the cold front, frequently from west to east; and (d) warm, dry air aloft, which overlies the warm, moist air. Extreme heating by the sun on a humid summer afternoon appears to help "trigger" the tornado, because most tornadoes occur in late afternoon or evening.

The storm is characterized by a funnel-shaped cloud that seems to form in the overlying chaotic cloud mass and grow downward to the ground. Frequently, the portion that touches the ground is only a few hundred yards in diameter, and consists of extremely violent winds (up to several hundred miles per hour) that pick up dust and debris, giving a visible funnel appearance. The storm generally moves in a northeasterly direction at a speed of 25 to 50 miles per hour (although there are frequent exceptions to this). The damage from the tornado comes mainly from three sources: the extremely high winds, which are strong enough to destroy most frame buildings; the sudden decrease in air pressure as the tornado passes overhead; and the drenching rain and hail that often accompany its passage.

Many investigators have attempted to study the tornado by using models and have succeeded in duplicating to some extent the appearance, if not the dynamics, of the natural storm. It is unlikely that one could actually create a tornado "in a box," but a model can be useful in examining the motions of air in a circular vortex. For example, what causes air to begin moving in a circular motion? Does the air in the vicinity of the whirling vortex get drawn into the vortex, or does it get pushed aside? Does the air within the vortex itself flow upward or downward? Why is a tornado vortex visible? Is it composed of water droplets, debris, dust, or ice crystals? The model can suggest some answers to these questions.

In an article entitled "A Tornado Model and the Fire Whirlwind" by James Miller in *Weatherwise,* the author reports some investigations with a simple vortex model. He believes that each microscopic droplet in the cloud formed in the vortex is acted upon by two forces. One is the *drag* of air spiraling toward the center. The other is the centrifugal reaction of the particles' rotation about the center. Inside the core of the vortex, the centrifugal reaction is stronger, and particles are thrown outward. Outside the cloud column, air drag is stronger, and particles are forced inward.

According to the model, the author hypothesizes two causes for a tornado:

1. Air must be drawn rapidly out of a region high above the earth's surface.
2. Fresh air coming in at lower levels must have some absolute rotation in the very beginning.

Since a tornado is a cyclonic storm with very low pressure at its center, air tends to rush toward the center from all sides and ascend rapidly inside the funnel. For this motion to continue, there must be some mechanism for this air to spread out or speed up and leave the vicinity when it reaches high levels. It is possible that the jet stream provides this mechanism. Recent studies on jet streams indicate that a region of divergence at high levels frequently accompanies the core of a jet stream, and this diverging air may provide the high-velocity winds that are apparently a necessary condition for tornado development.

The tornado model described here is patterned after one designed by Bellaire and Stohrer in 1963. As you construct it, keep in mind the following questions with the possibility of improving on the methods used here. Where does the tornado funnel begin—at the ground or high in the clouds? What produces the whirling action in a tornado? What is the usual direction of rotation?

Figure 35 illustrates the general appearance of the tornado chamber in action. The main chamber is an open-sided box with sliding glass panels that can be moved horizontally along the base to permit the entry of air from the sides. Smoke, admitted at two levels from a smoke generator, forms the visible vortex.

Construct the chamber from 4 corner posts of plywood,

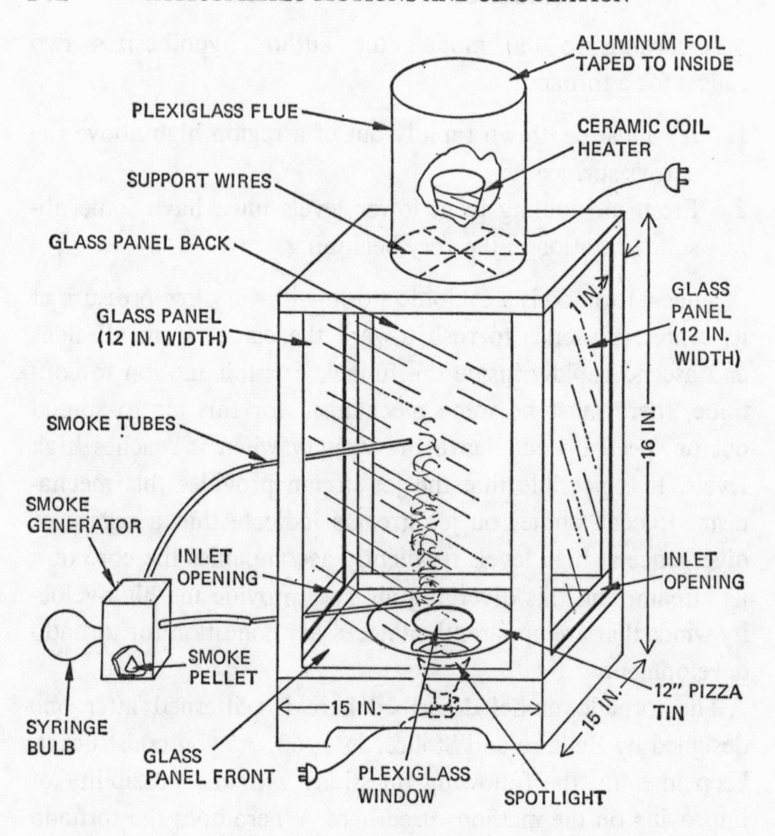

Fig. 35. Tornado Vortex Chamber

1 inch by 1 inch by 16 inches long, into which have been carved grooves running the length of each post on 2 adjacent sides. The grooves must be sized to enclose and support the glass or plexiglass panels that will form the sides of the chamber. One post, the one that will form the left front corner of the box, should be drilled with 2 holes at different levels large enough to accommodate 2 ⅛-inch or ¼-inch copper or steel tubes, which will admit the smoke. Nail the posts to a base of plywood, 15 inches square, into which a 10-inch hole has

been cut. The base must also have grooves along the sides to accommodate the sliding panels.

Prepare a 12-inch pizza or pie tin by cutting a 3-inch hole in the bottom and gluing over it a piece of plexiglass. This small, plexiglass window will be placed over a spotlight in the working model. Nail or tape the tin over the hole in the plywood base.

The panels are 12 inches by 16 inches long, about 1 inch narrower than the sides of the box, so that slots of variable width can be opened along either the right or left sides. You will probably find plexiglass an easier material than glass to work with for this purpose. Slide the panels into place through the top opening, and paint the back panel on the outside with black enamel, or tape a piece of black construction paper over it, to enhance the visual effect of the vortex.

The top of the chamber is a sheet of plywood, like the base 15 inches square with a 10-inch hole cut into it for the stack. Again, grooves along the bottom sides will provide paths for the sliding panels. The plywood top can be rested on the support posts and removed as desired for access to the chamber. The draft-producing stack is made of 2 12-inch-long plexiglass tubes, one 10 inches in diameter to fit in the hole in the chamber's top, the other just big enough to fit inside the first tube. The first tube is taped to the plywood; the second is raised or lowered within the main chamber to vary the effects of the vortex. Before fitting the tubes together, glue or staple coverings of aluminum foil to the inside of both. Rig a wire rack in the bottom of the inner tube to support a ceramic heater coil.

Insert the smoke tubes through the left corner post, and attach lengths of rubber tubing. When the chamber is functioning, you can either blow cigarette smoke directly through the tubes, or run the tubes into a small, smoke-generating box. This is simply a 3-inch-square plywood box, open at one end, with holes to admit the rubber tubes. A rubber bulb syringe can be inserted through a hole in the opposite side and pressed regularly to force the smoke out the tubes. Put a burning piece of cigarette or a smoke pellet in an empty pie tin, and place the box over it.

When the chamber is completely constructed, suspend it on bricks or boxes, allowing enough of an opening beneath for the spotlight. To operate the model, fill the flat pan at the base with ¼ inch of water, and turn on the electric heating coil in the base of the stack, and the spotlight below. The side panels should be adjusted to allow room for air to enter in such a direction as to give a rotary motion to the rising air within the stack. For counterclockwise rotation, allow the air to enter through a ½-to-1-inch vertical slit on the right side of each panel, by sliding the panels to the left a short distance. Different parts of the vortex can be investigated by admitting smoke through the smoke jets. Placing a beaker of dry ice and water in a corner of the chamber produces a dense white cloud that improves the visibility of the vortex.

As you work with your tornado model, try to answer some of the following questions:

1. Can you reverse the direction of rotation of the "tornado" funnel?
2. Does the funnel descend from above or rise from below?
3. What is the rate of temperature change with height (lapse rate) from bottom to top of the chamber? You will need a series of thermometers placed at strategic locations within the chamber to answer this question, or use the

thermopile described in the section on relative humidity, moving it back and forth and up and down.

4. Is a temperature inversion necessary to initiate the funnel or vortex? A temperature inversion refers to an increase of temperature with height.
5. Is this model more analogous to a waterspout or a tornado over land? In a waterspout, water drops and cloud droplets intermingle to form a visible funnel.
6. Is there a downdraft or an updraft in the center of the "tornado" funnel?
7. What is the pressure gradient from the outside of the funnel to the center? How could it be measured?
8. In what ways does the model resemble the real tornado? In what ways is it different? What about the electrical effects in a tornado? Can these be simulated? Recent theories tend toward an electrical explanation for tornadic winds. Can these be investigated?
9. Suppose a high electrical potential were placed between the pan of water and the discharge stack; for example, 200 volts D.C. at very low amperage, such as can be obtained from certain types of batteries. What would be the result? Try a potential of 100,000 volts using an electrostatic generator such as a Wimshurst static machine or Van de Graaff generator. Observe the results. Do some additional reading on the dynamics of dust whirls, tornadoes, and waterspouts in order to obtain ideas for modifying your vortex chamber and devising additional experiments.

INVESTIGATION 18. (E)

Measuring Surface Air Movement

MATERIALS NEEDED:

Transparent plastic soda straw, ¼-inch diameter
Small styrofoam ball, 3/16-inch diameter
Stiff cardboard or balsa wood, 3 inches by 12 inches by ¼ inch
Modeling clay
2 pins

Besides rain, snow, and violent storms, the meteorological phenomenon that people are probably most conscious of is moving air. A sudden high wind can ruin a picnic, and winds over lakes and bays can generate unpleasant waves in a very short time. Even cars are not immune to sudden gusts of wind on the highway, particularly if moving at right angles to the direction of the wind.

A simple device for measuring wind speed can be made from a transparent soda straw, a very small styrofoam ball that will fit inside the straw, some cardboard or balsa wood, and pins. Styrofoam balls can be made from any styrofoam product, such as insulated cups, ice coolers, or packing materials. The soda straw must be perfectly clean and dry; use pipe cleaners to keep the inside of the straw free of lint. With modeling clay, close off the bottom of the soda straw. Make a tiny pinhole in the side of the straw near the top for the purpose of calibration, and cut out a notch near the base to allow wind to blow into the straw. A paper scale with high- and low-range markings on it can be taped to the straw, and the completed assembly attached to a balsa wood or cardboard base with pins, as shown in Figure 36.

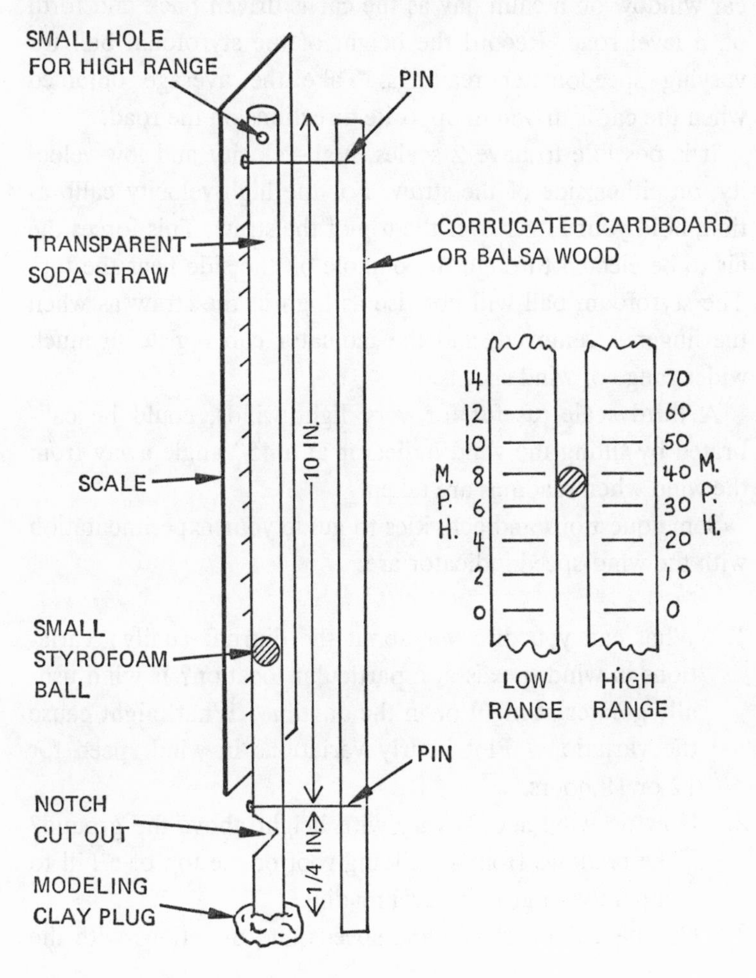

Fig. 36. Wind-Speed Indicator

To operate the wind-speed indicator, hold the device at arm's length with the open notch at the bottom of the straw facing into the wind. Air moving into the notch will lift the styrofoam ball, which should fit loosely inside the transparent straw, to a height that is proportional to the wind speed.

To calibrate your wind-speed indicator, hold it outside a car window on a calm day as the car is driven back and forth on a level road. Record the height of the styrofoam ball for varying speedometer readings. Take the average obtained when the car is driven in opposite directions on the road.

It is possible to have 2 scales, high-velocity and low-velocity, on either side of the straw. For the high-velocity calibration, hold your finger over the top of the straw. This forces the air to be ejected through the pinhole on the side near the top. The styrofoam ball will not rise as high in the straw as when the finger is removed, and the indicator can register a much wider range of wind speeds.

A third scale, useful for very light winds, could be calibrated by tilting the wind indicator at a 45° angle away from the wind when readings are taken.

Some questions and activities to guide your experimentation with the wind-speed indicator are:

1. What can you discover about the diurnal (daily) variations in wind speeds at a particular location? Is wind usually greater at night or in the daytime? What might cause the variations? Plot hourly variations in wind speed for 12 or 18 hours.
2. How do wind speeds vary with height above the ground? Take readings from a building roof or the top of a hill to compare with ground-level readings.
3. Compare the surface-wind speeds and directions with the

winds aloft as indicated by cloud movements. (See Investigation 5 for methods of measuring cloud directions and speeds.)

4. Can you measure wind gusts with your indicator? What is the maximum gust speed you can record? How long does it take to reach maximum gust speed?
5. What is the maximum period of prolonged high wind? In summer? In winter?

INVESTIGATION 19. (E)

Constructing a Sensitive Anemometer

MATERIALS NEEDED:

2 plywood boards, ¼ inch by 6 inches by 12 inches
4 small wood screws, ¾-inch long
Spring-type clothespin
Cigarette

An ultrasensitive wind-speed indicator for measuring very light air movements in rooms, corridors, caves, or outdoors on nearly calm days can be made by using a smoke source, a suitable vertical scale for estimating the speed of the rising smoke, and a horizontal scale to register the sideways drift of the smoke in the same time interval.

A diagram of the device is shown in Figure 37. Using small screws, fasten 2 boards, each ½ inch by 6 inches by 12 inches, together at right angles to each other. Draw horizontal and vertical scales on the vertical board as shown in the diagram. The graduations should be made clearly visible, using black or white paint, or a felt-tipped pen. A clothespin holder can be

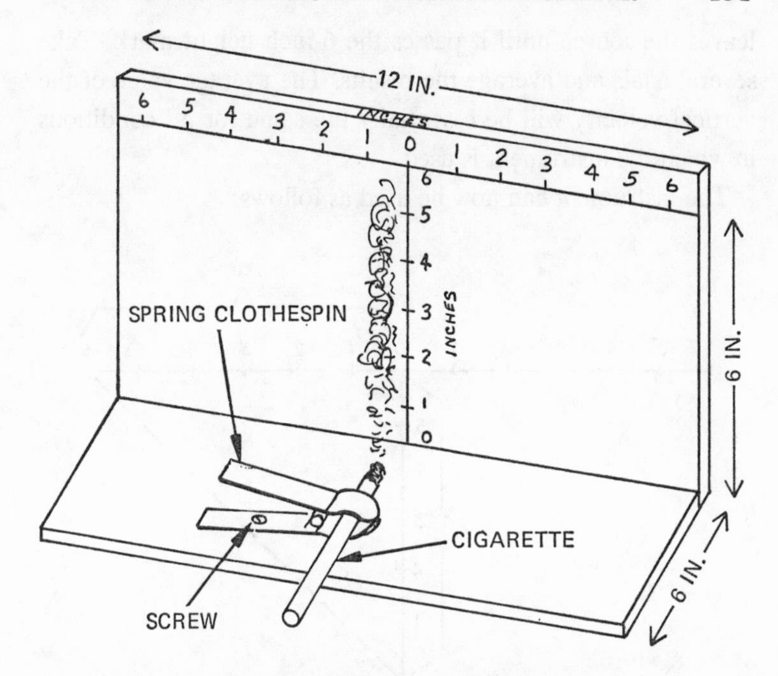

Fig. 37. Sensitive Anemometer

attached to the base with a small screw to hold the smoke source, such as a burning cigarette or punk.

To use the indicator, you must know the rate at which smoke rises from the smoke source. Once this is obtained, it can be assumed to be a fairly constant rate for most conditions. The speed of horizontal air movements can then be calculated by observing the smoke drift on the horizontal scale and using a trigonometric relation involving the rate of vertical motion and the angle of drift.

Determine the rate by measuring the velocity of the rising smoke in inches per second. To do this, time the rising smoke, under calm conditions, with a stop watch or, working with a companion, a watch with a second hand, from the instant it leaves the source until it passes the 6-inch height mark. Take several trials and average the results. The average value of the vertical velocity will be essentially the same for all conditions in which the instrument is used.

The instrument can now be used as follows:

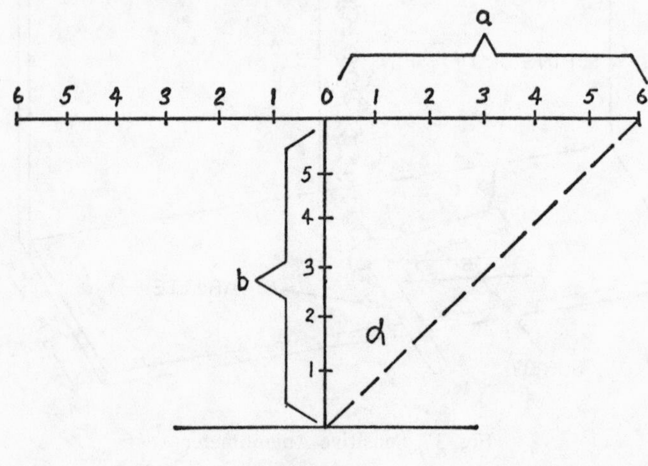

Fig. 38.

From the diagram, $\tan \alpha = \frac{a}{b}$,

Let V=horizontal velocity of the wind and Z=vertical velocity of the rising smoke.

Then $\tan \alpha = \frac{V}{Z}$, $V = Z \tan \alpha$

Substituting for $\tan \alpha$ from above, $V = Z(\frac{a}{b})$

As an example, suppose Z averages 2 inches per second. On a given measurement, imagine that the breeze causes the smoke column to drift horizontally so that it crosses the horizontal scale at a point where a=3 inches. The height "b," of course, is 6 inches. Using the equation $V=Z(\frac{a}{b})$ from above, the calculation is:

V=(2 inches per second) (3 inches/6 inches)=1 inch per second

To simplify the measurements and enable you to take several measurements quickly, a chart similar to the following might be prepared:
(For b=6 inches, Z=2 inches per second)

a (inches)	*V (inches per second)*
1	1/3
2	2/3
3	1
4	4/3
5	5/3
6	2

You will find that your instrument is extremely sensitive to very light air currents, and the smoke plume will waver quite a bit due to turbulence. For this reason, all measurements will need to be taken several times and the results averaged to get meaningful data.

Can you think of any additional uses for this instrument? On a calm day, can you measure small air currents at the edges of a shade tree? Near an asphalt road or parking lot? Near lake shores, buildings, or hilly areas? Can you detect the beginnings of convection currents over plowed fields? Convection, you will recall, is the vertical motion that results when there are differences in density in the atmosphere; convection currents start when air is heated from below. Can you identify a pattern in wind behavior just as a cloud passes in front of the sun?

For measuring vertical air movements near the ground or

turbulence effects near buildings and trees, you might try to invent an instrument using soap bubbles. How would you obtain quantitative information using soap bubbles? What is their rate of ascent? Do they follow wind currents faithfully? To answer these questions, you will need to develop a reliable soap-bubble generator, determine the rate at which soap bubbles rise, and devise a way for measuring their horizontal motion. There is a real challenge in building an instrument of this kind!

INVESTIGATION 20. **(M)**

Balloon Tracking of Low-Altitude Winds

MATERIALS NEEDED:

Plastic dry-cleaning bag
2 balsa wood strips, 1 inch by ¼ inch by 16 inches
Thin cardboard disc, 6-inch diameter
14 small birthday candles, 2 inches long
Tape, masking or Scotch
Binoculars with tripod mount
2 large protractors
2 metal strips, ⅛-inch thickness, with ⅛-inch holes drilled in each end
Several sheets of polar-co-ordinate graph paper

The balloon-tracking way of measuring winds aloft at relatively low altitudes (1,000 to 5,000 feet) is patterned after a method used by the U. S. Weather Bureau. A "pilot balloon" is improvised from a plastic dry-cleaning bag and filled with helium gas or hot air as the lifting agent. Helium gas can usually be obtained from a college chemistry laboratory, if there is one in your vicinity. If not, small pressure tanks of helium gas can be purchased from most laboratory supply companies. Once the balloon is ready for takeoff, the principles involved in measuring wind speed are very much the same as those used in the smoke anemometer of the preceding investigation.

The plastic bag should be carefully sealed with adhesive tape if helium is to be used. After the bag has been sealed at the top and filled with gas, close the bottom tightly with a rubber band, and wind a length of flexible wire around the neck, allowing the excess wire to serve as a weight hanger (see Figure 39). From this hook, suspend paper clips or metal wash-

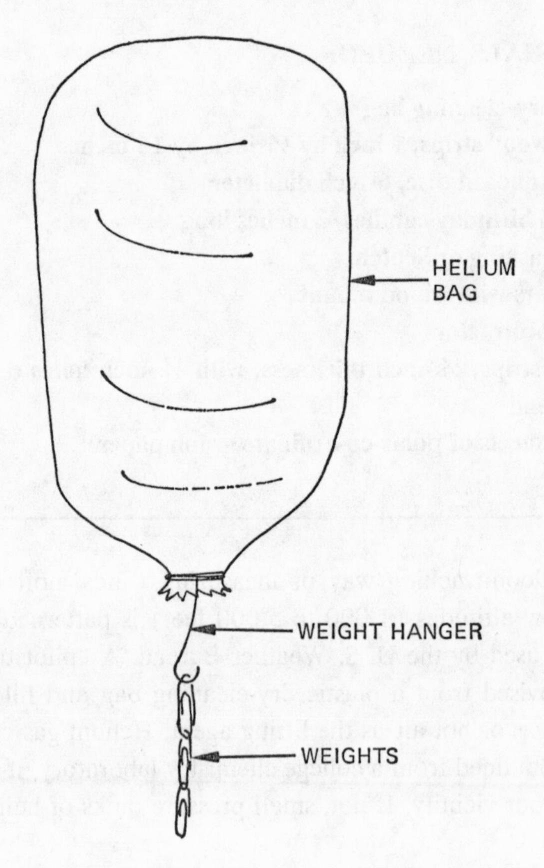

Fig. 39. Helium Balloon for Tracking Low-Altitude Winds

ers until the balloon is exactly in balance, that is, neither rising nor sinking.

When the balloon is in balance, remove the smallest possible weight so the balloon will rise at a constant rate. By tying a fixed length of cord, say five feet, to the neck of the balloon and timing how long it takes the balloon in calm air to reach the end of the line, you will have the ascent rate to be used in the following calculations. Take several trials and average the results.

An alternate device, using hot air as the lifting agent, can be constructed as shown in Figure 40. In this case, the plastic

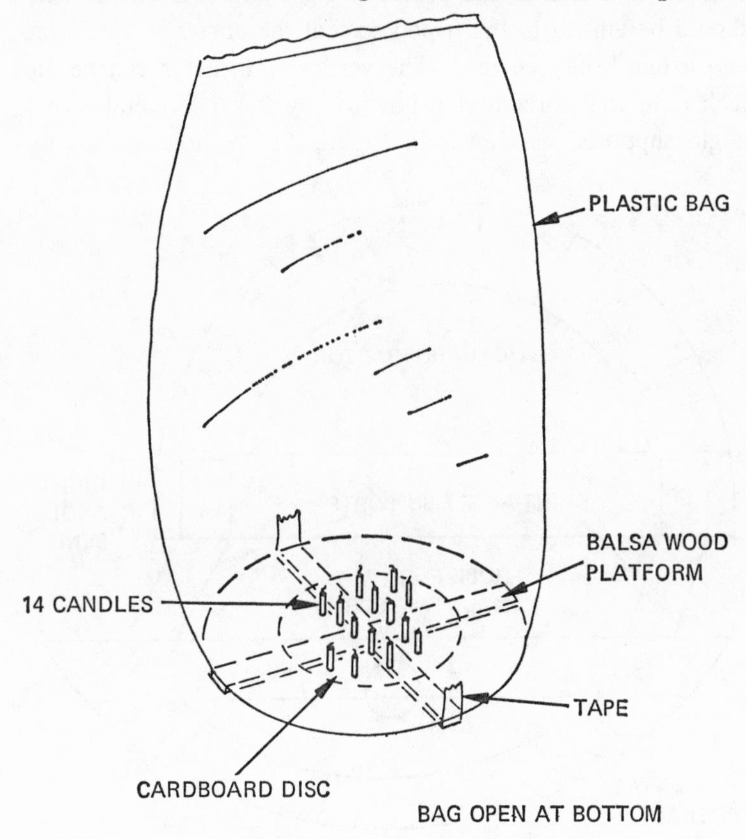

Fig. 40. Hot-Air Balloon for Tracking Low-Altitude Winds

dry-cleaning bag is sealed at the top but left open at the bottom end. A platform, made of cardboard and pieces of balsa wood, is taped to the sides of the bag as shown in the diagram. Small candles, such as birthday candles, are attached with wax to the platform and provide the heat source for heating the air inside the balloon. The rate of ascent with this device will, in all likelihood, be more variable than with the helium balloon. Measure the ascent rate over several trials, as before.

To track the balloon, you will need a pair of binoculars mounted on a camera tripod. Two large protractors, at least 6 inches in diameter and placed at right angles to each other, should be affixed to the tripod so that the angles of elevation and azimuth can be read. The vertical protractor can be attached to the horizontal protractor by 2 small metal right-angle supports, as shown in Figure 41. A hole should be

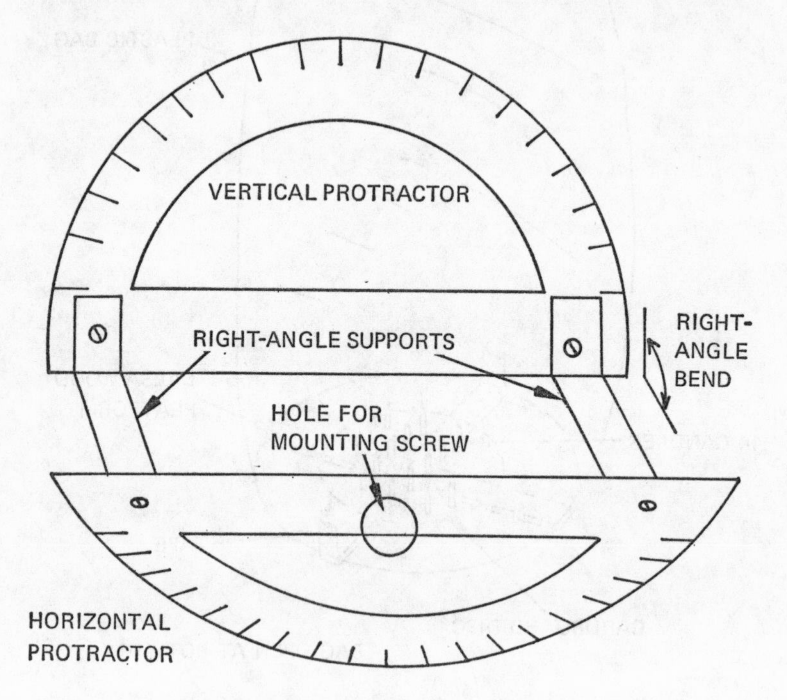

Fig. 41. Suggested Method of Mounting Protractors

drilled in the center of the straight side of the horizontal protractor. The mounting screw for the binoculars goes through this hole to keep the protractors in position. Figure 42 illus-

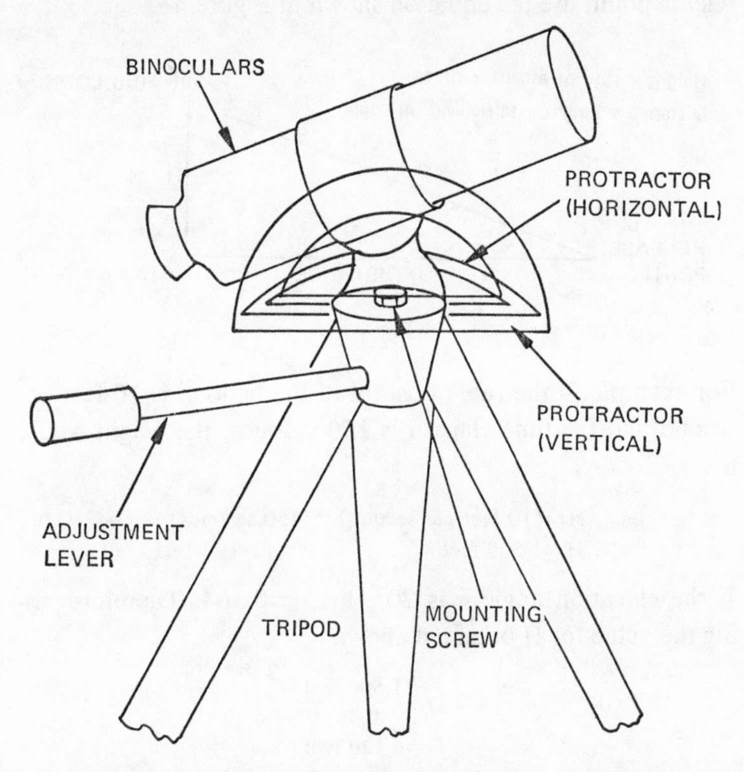

Fig. 42. Tracking Instrument for Low-Altitude Winds

trates the complete arrangement of materials for the tracking instrument.

When all is in readiness, secure the help of an assistant who will release the balloon and give you time signals for taking readings of the azimuth and elevation angles. Track the balloon with the binoculars from the moment of release. Obtain readings of azimuth and elevation angles every 30 seconds until the balloon disappears; have your assistant record your readings.

To compute the height and distance of the balloon from the release point, use the equation shown in Figure 43.

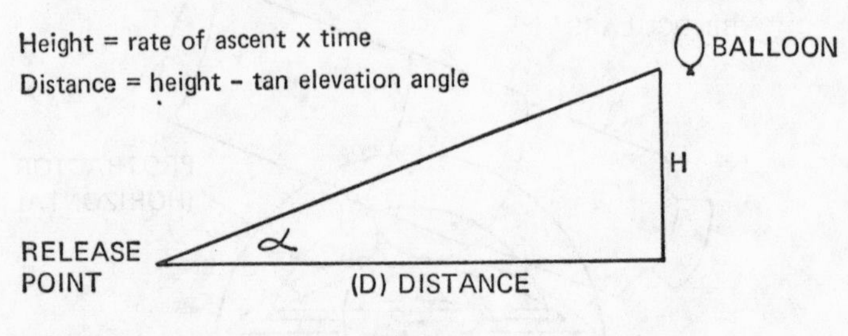

Fig. 43.

For example, if the rate of ascent of the balloon is 10 feet per second and the time elapsed is 150 seconds, the height would be:

$$H = (10 \text{ feet per second}) \times (150 \text{ seconds})$$
$$H = 1{,}500 \text{ feet}$$

If the elevation angle α is 20°, tan $\alpha = 0.364$. Therefore, using the value for H obtained above:

$$D = \frac{1{,}500 \text{ feet}}{0.364}$$
$$D = 4{,}120 \text{ feet}$$

Using the elevation angles, rate of ascent distance, and the elapsed time, calculate wind velocity. To obtain the average wind speed between each observation, subtract from each distance D, the distance traveled by the balloon for the previous observation. This gives the distance traveled during the interval between observations. Divide this distance by the time interval, 30 seconds, to get the average wind speed for the interval.

A table recording all these values might look like this:

	Elapsed Time (t)	Elevation Angle (α)	Azimuth Angle (β)	Tan α	Height (H)	Distance (D)	Average Wind Velocity (V) Between Points
Trial	t (sec)	α°	β°	Tan α	H (feet)	D (feet)	V (ft/sec)
							11.9
1	30	40	90	.839	300	357	
							16.7
2	60	35	95	.700	600	857	
							23.4
3	90	30	100	.577	900	1560	
							33.6
4	120	25	105	.466	1200	2575	
							51.5
5	150	20	110	.364	1500	4120	

Ascent Rate = 10 ft./sec.

To determine wind directions aloft, plot the information from the azimuth angles obtained at each reading and the calculated distance of the balloon from the release point. Use polar co-ordinate paper, assuming that the release point is located at the center of the graph. A diagram of this plotting might look like Figure 44.

Locate for each observation the intersection of the total horizontal distance, D, traveled by the balloon from the point of origin, and the azimuth angle obtained for that trial. Connecting the intersections with straight lines on the polar co-ordinate paper shows the path of the balloon and the distance traveled between each observation.

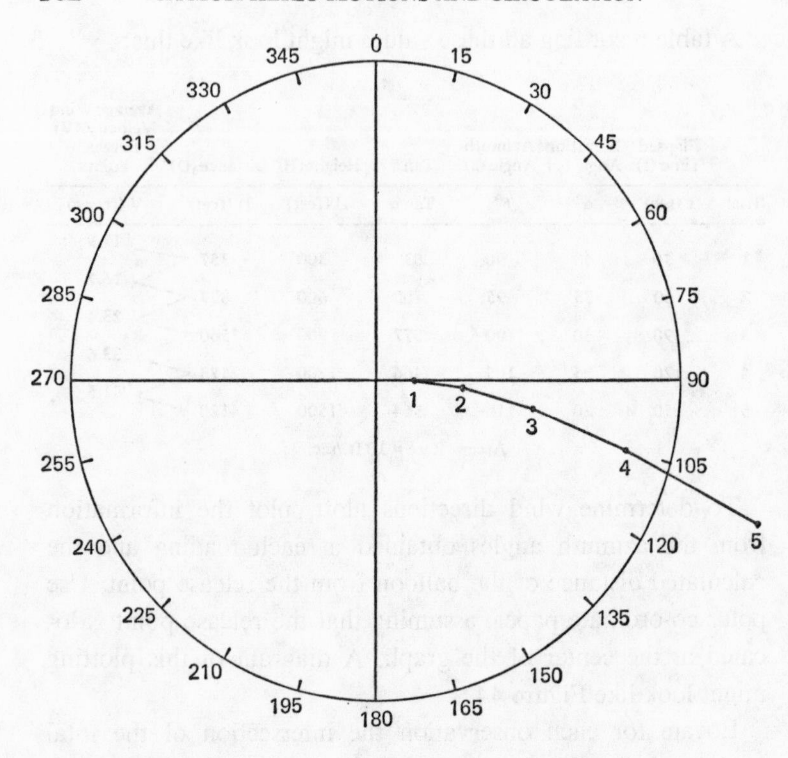

Fig. 44. Sample Plot of Balloon Flight

Questions you might attempt to answer as you do this experiment are:

1. How do winds vary with height?
2. What is the height of the balloon at the time of its disappearance? What causes the balloon to disappear—clouds or lack of contrast with the background sky? How could the latter problem be somewhat alleviated?
3. What is the height of low clouds on the day of your experiment? Determine the height at which the balloon disappears in the clouds. How does the height obtained

by this method compare with that obtained by the triangulation method described earlier in this book in the section on measuring cloud heights?

4. How can you compute the rate of ascent of the balloon experimentally? You might use the method for measuring the heights of clouds to determine this value.
5. Compute the wind shear between the ground level and the level of the last observation. Wind shear is the difference in wind speed divided by the distance over which this difference is measured. For example, if the wind speed at a height of 600 feet is 14 feet per second and at 1,200 feet it is 21 feet per second, the wind shear is

$$\frac{(21\text{–}14) \text{ feet per second}}{(1{,}200\text{–}600) \text{ feet}}$$

$$= \frac{7}{600} \text{ per second} = .011 \text{ in units of } 1/\text{seconds (reciprocal seconds)}$$

For purposes of comparison, you might be interested to know that the vertical wind shear between the base and the center of a jet stream is approximately (.005–.010) per second.

6. Where are wind shears the greatest—near lakes, over plains, over wooded areas?
7. How does the balloon-ascent rate vary with air temperature? With volume of the balloon? (Use a larger plastic bag to increase volume.)
8. Study smoke plumes from chimneys. Can these be used to gauge low-altitude winds?
9. Being very careful, you might try using the gas from a propane torch to inflate your balloon. How does its ascent rate compare with that of a helium balloon? For reasons of safety, fill the balloon outdoors using a long connecting nose from a gas outlet. *Be sure there are no open flames or burning cigarettes in the vicinity.*

INVESTIGATION 21. (M)

Model of Frontal Surfaces

MATERIALS NEEDED:

Plate glass or plexiglass, 2 inches by 8 inches by ¼ inch
20-gauge angle wire (iron or aluminum), ⅛ inch by ¾ inch by ¾ inch. Total length needed: 42 feet, 8 inches
Putty (glaziers)
Plastic guides, ⅛ inch; 6 feet, 8 inches total length
2 aluminum sheets, .020 inch by 2 feet by 8 inches
Salt
Food coloring

While listening to weather reports or watching them on television, you have heard the weatherman describe various kinds of weather "fronts." Perhaps you heard him say that a "cold front passed Minneapolis during the night" or that "an approaching warm front is bringing fog and rain to the area."

What are "fronts"? Why do they cause changes in the weather? Do they just occur on the ground, or do fronts have a three-dimensional structure?

To obtain a better idea of what weather fronts are, and to study their action on a small scale, construct a model using various densities of colored water. (To alter the density of the water use varying concentrations of salt.) The different densities of water will simulate air of different temperatures, and the interaction of the various water layers will duplicate, to some degree, what takes place in the real atmosphere when air masses of radically different temperatures and densities clash with one another.

The basic process underlying the movement of weather fronts is the interaction that occurs when a cold, dry air mass encounters a warm, moist air mass. The former, being more dense, tends to flow underneath the latter and cause the warm, moist air to be lifted slowly off the ground. This lifting action generally produces cooling of the warmer air, condensation, rain, and stormy conditions in the region of the front. In the model, very salty water is used to represent the cold air mass, less salty water represents the warm air mass, and fresh water represents an atmosphere of moderate temperature separating the 2 advancing air masses.

To build the model shown in Figure 45, construct a frame

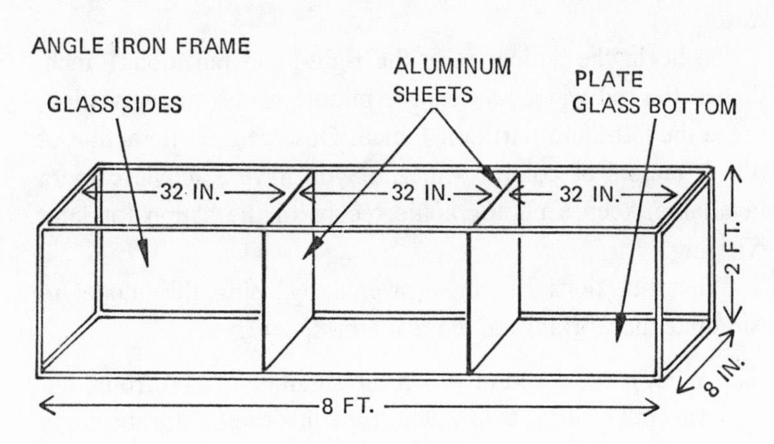

Fig. 45. Density Channel for Simulating Weather Fronts

8 feet by 2 feet by 8 inches, using 20-gauge angle wire with dimensions of ⅛ inch by ¾ inch by ¾ inch. Use ¼-inch clear glass for the sides of the density channel, sealing the glass to the frame with putty; the bottom of the tank is a single sheet of plate glass. The two partitions are made from .020-inch-thick aluminum sheets, fitted into ⅛-inch plastic or glass guides cemented to the sides and bottom of the channel. All the necessary materials for building the density channel can be obtained from a school industrial arts or metal shop. Such a shop will also have the necessary tools for working the materials and constructing the apparatus. When finished, paint all joints with white, waterproof paint.

Fill the center section of the channel with fresh water to approximately an 18-inch depth. Completely fill the left-hand section with water, to which 2 generous tablespoonsful of salt have been added. Stir blue food coloring into this water. Completely fill the right-hand section with water to which 1 tablespoonful of salt has been added, and stir red food coloring into it. All the water you use should be at room temperature.

To begin the action, raise the right-hand partition 1 inch. When the red water reaches the middle of the center section, raise the left-hand partition 1 inch. Observe the interaction of the 2 masses of colored water. If you have a movie camera available, keep a photographic record of the action for later viewing.

What questions can be answered by using this model to simulate the workings of the real atmosphere?

1. What is the speed at which the simulated "cold front" advances? Using a stop watch, time this rate of advancement during several trials. How does this compare with the

average speed of an atmospheric cold front, 30 to 35 miles per hour?

2. What is the slope of the advancing wedge of the "cold front"? In the atmosphere, the slope is usually about 1 to 50 or 1 to 100, or 1 foot of vertical height for each 50 or 100 feet of horizontal distance.
3. What finally happens to the simulated "warm air mass"? Does it mix with the "cold air"? Does it demonstrate turbulence and eddies as it is pushed upward?

Remember that this is only a demonstration model, and many of the factors operating here will be quite different from those in the real atmosphere. For one thing, water is about 750 times as dense as air. The vertical scale in your model is exaggerated about 20 times over that in the real atmosphere. The elapsed time in the model is about 1/10,000 the time it actually takes for corresponding movements to happen in the real atmosphere. Considering all of these differences, can you learn anything about the real factors operating in the atmosphere when cold and warm air masses meet? As in all models, the advantages of the density channel are visibility, small size, and the ability to repeat and duplicate certain actions for further study.

Chapter 5

ATMOSPHERIC OPTICS

Atmospheric optics refers to the study of those conditions of the atmosphere that can be explained by changes in light and color. Rainbows and halos, for example, come under this heading, as do mirages, colored sunsets, and changes in sky color.

The constantly changing appearance of the sky and cloud cover can be an endless source of interest to the meteorologist and to the non-scientist, as well. During one summer on Guam, I was fascinated by the changing colors of the water surrounding the island, from dark gray to deep, bright blue. By correlating these changes with the degree of cloudiness, I found a direct relation between the shading of the water and the fraction of sky cover: the clearer the sky, the bluer the water.

To use another personal illustration, I have often been intrigued by the contrast of the brilliant blue sky over the high mountains of Colorado and the rather washed-out blue generally visible from sea-level elevations. Those sapphire blue skies undoubtedly add to the delight of a Colorado skiing holiday. But what causes these startling color differences? Is it the clarity of the air, the lack of dust or water vapor, or the difference in air density itself? Perhaps we can use models, once again, to suggest answers to some of these questions.

INVESTIGATION 22. (M)

Model of the Sky–Atmospheric Scattering

MATERIALS NEEDED:

1 rectangular aquarium, glass, 5-gallon capacity
1 flashlight
1 pair of polaroid sunglasses or polaroid filter
10 grams sodium thiosulfate
100 milliliters of concentrated sulfuric acid

The changing color of the sky has long been considered a significant factor for weather prediction. Just a few seconds can cause startling shifts in the hues and shadings of the sky at sunrise or sunset, and a very old saying refers to these changes:

Red sky in the morning,
Sailors take warning;
Red sky at night,
Sailor's delight.

This observation was originally made by Theophrastus, Aristotle's pupil, and was included in his *Book of Signs,* written about 300 B.C. Theophrastus observed that a red sunrise and a halo around the sun or moon portended rain, while a red sunset portended good weather. One recent study in meteorology has shown this statement to be true about 70 per cent of the time. Perhaps the accuracy of the saying is based on the fact that halos or rings around the sun or moon indicate the presence of ice crystals high in the air, which often appear in upper air currents in advance of storms. Red sunsets, on the other hand, are usually caused by dust and haze in the lower atmosphere, frequently more abundant at the end of the day than at the beginning.

A partial understanding of what causes such color changes can be aided by a model made of water and a few common chemicals. You will need an aquarium (about 5-gallon size), a flashlight, a pair of polaroid sunglasses or a polaroid filter, about 10 grams of sodium thiosulfate, and a small bottle of concentrated sulfuric acid. Fill the aquarium with water to within a few inches of the top, and mount the flashlight near one end of the tank with its light beam directed horizontally through the water.

The flashlight, as you might guess, simulates the sun shining through the atmosphere—in this case, water. At the start of the observation period, the water is clear—the atmosphere has a minimum of particles suspended in it. Under these conditions, the flashlight "sun," when viewed through the full length of the water-filled aquarium, appears white or only slightly yellowish. This is analogous to the situation of an observer looking at the sun from a point high in the atmosphere, above the region where particles are found in abundance.

To reproduce atmospheric conditions nearer the ground,

about 10 grams of sodium thiosulfate should be stirred into the clear water. What effect, if any, does this have on the appearance of the "sky" or the "sunlight" passing through it?

After the sodium thiosulfate has completely dissolved, add about 5 drops of concentrated sulfuric acid to the water, and stir thoroughly. After a minute or two, a noticeable change in the appearance of the water will occur. Observe the appearance of the flashlight beam as it passes through the water, the color of the water itself, and the lighted end of the flashlight when viewed through the full length of the aquarium. The effects will be heightened by darkening the room.

The chemical reaction that occurs between the sodium thiosulfate and the concentrated sulfuric acid produces free particles of sulfur. With time, these small particles grow by accretion until they are large enough to affect the light passing through the solution. Scattering, or multiple reflections, of the various components of white light occurs, depending on the wavelengths of the light. Short wavelengths are usually scattered more than long wavelengths. The following questions may enable you to discover for yourself what caused the progressive changes you observed in the solution, and may suggest a possible explanation for similar effects in the atmosphere.

1. What colors are contained in white light?
2. How do wavelengths of light vary with color? Which color has the longest wavelength? The shortest wavelength?
3. Which wavelengths (colors) are scattered more by small particles? By large particles?
4. When blue light is scattered or removed from white light, what color of light remains? If you consider white

light to be made up of a composite of red, orange, yellow, green, blue, indigo, and violet light (Sir Isaac Newton's discovery), what color would remain if all wavelengths shorter than green—i.e., blue, indigo, and violet—were removed?

5. Why does the "sky" (or water in the aquarium) look bluish at one point in the sequence of particle growth? What color would the water appear if the particles scattered blue light the most?
6. Why does the "sky" (or water in the aquarium) become whitish as suspended particles grow larger? (What color will the eye perceive when all colors [wavelengths] are scattered and come to the eye simultaneously?)
7. Why does the "sun" (or light source) seen through the water appear more reddish as particles grow larger? Remember, long wavelengths can penetrate particle-laden atmosphere better than short wavelengths.
8. From the side, observe the water through a polaroid filter. Such a filter permits light waves of only one plane of vibration to pass through. For example, light vibrating in a vertical plane (up and down) would pass through a polaroid filter only if the internal orientation of the filter permitted light of vertical vibrations to pass. Light of any other plane of vibration would be blocked or absorbed. Turn the filter through 90°. In what direction is scattered light from the aquarium polarized—horizontally, vertically, or in some other direction? How could you determine this direction of polarization? Could you determine the direction of polarization of scattered light in the atmosphere by the same method?
9. Observe the light that has passed from the source directly through the aquarium, by means of the polaroid

filter. Turn the filter through 90°. Through what angle is the scattered light rotated by the sulfur particles?

10. Why is the sky bluer at some times than at others? Would the wavelength (color) of the scattered light make a difference?
11. Why does the color of the sky generally become more white or grayish toward late afternoon? (How would increasing numbers of particles or introduction of larger particles to the atmosphere affect scattering?)
12. Why do the clouds in the west frequently become pink or reddish at sundown? If long wavelengths are scattered more in late afternoon, due to more large particles in the atmosphere, what colors would be predominantly reflected from the clouds?
13. For an observer on the surface of the moon, what color does the sky appear? What color does the sun appear to be? Did any stage in the experimental model simulate these conditions?

INVESTIGATION 23. (M)

Studying Hot-Plate Mirages

MATERIALS NEEDED:

1 electric hot plate or electric flatiron
1 slide projector
1 small square of cardboard, 12 inches square

Needed for suggested additional experiments:

1 electric fan
1 flat-bottomed metal pan
Ice cubes
1 aquarium

Most of us have had the experience of driving along a road on a hot summer day and seeing, suddenly, a mirage—"water" on the road ahead, although it hasn't rained for days. Have you noticed that the "water" seems to disappear as you get closer to it?

To gain a better understanding of what causes an atmospheric mirage, you can construct a model consisting of a hot plate or electric flatiron, a slide projector, and a small piece of cardboard. With these items, you will be able to study the bending or refraction of light beams as they pass through the air above a heated surface.

Arrange the hot plate or iron so that the projected light beam shines horizontally across the face of the hot surface. Place the projector about 1 foot from the hot plate and mount or hold the edge of the cardboard so that only a shallow beam of light passes above the heated surface. A beam about ⅛ inch high is sufficient. If the projector is focused on the far wall or a projecting screen, you should see the shadows of the plate and cardboard in sharp detail, with a bright band of light between them.

When your equipment is assembled, darken the room. For best results, have all doors and windows closed to avoid horizontal drafts. You will now see shadowy streaks moving and rising from the hot surface. These shadows indicate that differences in air density exist, giving rise to vertical convection currents above the heated surface. There may also be a marked "boiling" appearance near the bottom of the lighted image.

In both this experiment and the water mirage on the road, light rays are bent or refracted as they pass through strongly heated air. When light travels from one medium to another of different density, such as from cool to hot air, at an angle different from 90°, it "refracts" or changes its direction. Light travels faster in hot air than in cool air and consequently bends in an upward direction. In addition to simple refraction, the phenomenon of "total internal reflection," in which the light rays actually reflect from the interface between the cool and warm air, takes place. Figure 46 illustrates the principles involved in total internal reflection.

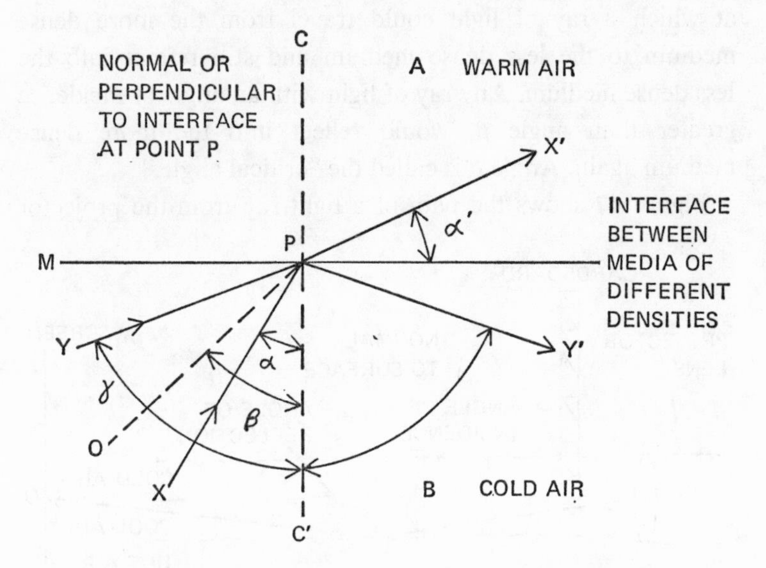

Fig. 46. The Principles of Total Internal Reflection

Warm air, A, the less dense medium, lies above cold air, B. MN represents the interface between these 2 air masses of different density. CPC′ is the normal, or perpendicular, line to the interface at point P.

A ray of light, XPX′, travels from the more dense to the less dense medium and refracts away from the normal as shown. The angle of incidence, δ, is smaller than the angle of refraction, δ′. Another ray of light, YPY′, travels up to the interface at point P, but instead of entering the less dense medium, A, it reflects downward into the more dense medium, B, again, because the angle of incidence, δ, is too large for refraction to occur at the interface. This process is called "total internal reflection." The angle of incidence, δ, is equal to the angle of reflection, δ′.

The angle, β, represents the maximum angle of incidence

at which a ray of light could travel from the more dense medium to the less dense medium and still refract into the less dense medium. Any ray of light with an angle of incidence greater than angle β, would reflect into the more dense medium again. Angle β is called the "critical angle."

Figure 47 shows the path of a light ray from the projector

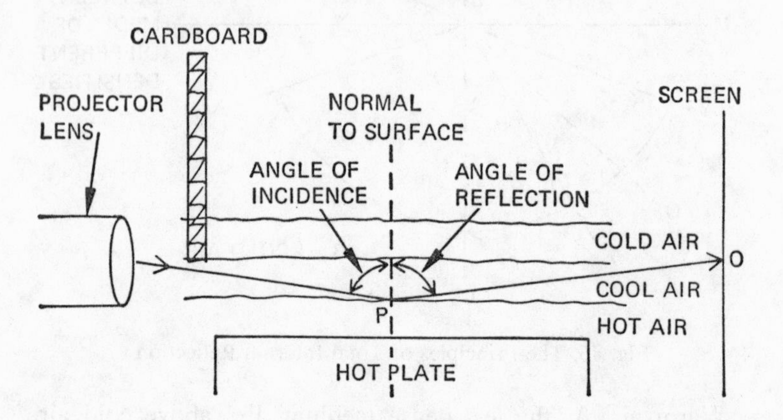

Fig. 47. Total Internal Reflection as Light Ray Passes from Cool Air (more dense) to Hot Air (less dense)

undergoing total internal reflection and producing a shimmering image on the screen. As the light ray approaches the surface of the hot plate or electric iron, it encounters air that is getting progressively hotter and hotter. Since light that travels from a more dense to a less dense medium bends away from the normal (perpendicular), the ray becomes more nearly parallel to the heated surface as it gradually is refracted while passing into warmer and warmer air. At point P in the diagram, the light ray reaches the critical angle of incidence, and is turned back into the denser medium and reflected to point O on the screen.

Because the layers of hot and cold air above the hot plate are constantly changing due to convection, the image on the screen gives a "boiling" appearance, comparable to the "twinkling" of stars or the "shimmering" of distant objects near the horizon in bright sunlight.

Figure 48 shows how the water mirage on the road forms

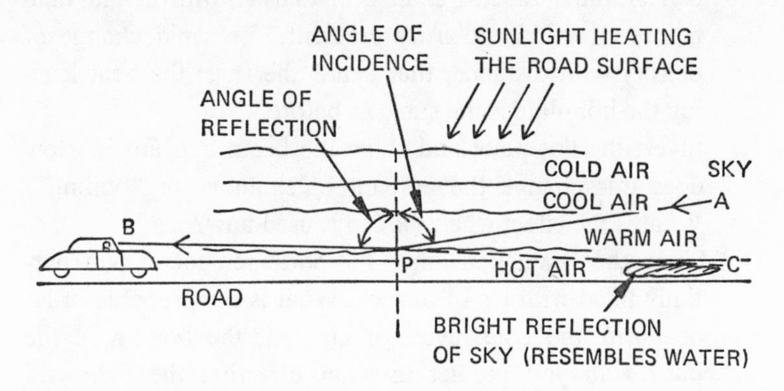

Fig. 48. Diagram of Formation of the Mirage, "Water" on the Road

in a manner similar to the hot-plate mirage. In this case, nearly horizontal light rays pass through successive layers of warmer air until they strike the hottest air, which is directly above the heated road surface. Due to total internal reflection, the light rays are bent upward along the line APB, entering the eye of the observer at B. Because he assumes the ray of light has been traveling in a straight line, it appears to him the light originated at point C on the road surface. Thus he sees on the road ahead a bright area (light from the sky) that looks exactly like a puddle of water.

Here are some additional experiments you can do to study the principles of light refraction further:

1. To show that it is the pattern of changing air density that bends the light, hold an electric fan so that it blows

down onto the hot surface. With a powerful fan, the bending seen on the screen or wall may be more than doubled. The region of mixing of warm and cool air is now closer to the hot surface, and the result is a more concentrated rate of change of density with height than before. It is this "density gradient," or rapid change of density with distance, that bends the rays; the heat leaving the hot plate is the same as before.

2. Invert the hot plate and shine the beam beneath it. How does this change the amount of shimmer or "boiling"? What is the effect when the fan is used now?
3. Shine the beam beneath a flat-bottomed metal pan partially filled with ice or dry ice. What is the probable order of warm and cold layers of air near the bottom of the pan? Can you predict in what direction the light will bend now? Are the results the same as before?
4. You might try to obtain density gradients without a heat source by mixing solutions in a flat-walled glass container, such as an aquarium. Fill the aquarium with water and let it come to rest. Shine the light through the tank close to the bottom, again using the cardboard to obtain a shallow beam. Quickly dump in a cupful of sugar or salt. Stir the solid a little to get it to dissolve near the bottom of the tank. What happens to the image formed by the beam of light projected through the solution onto a nearby screen or wall? Can you detect density gradients? Does the light bend upward or downward in this case? Try drawing a ray diagram to explain how the bending takes place.

Mirages formed by density gradients in the air are common in desert regions during the warmer hours of the day. The images of distant objects or the sky appear as inverted reflections below the objects, and are therefore called "inferior mirages." This type of image is particularly common on the west coast of Great Salt Lake.

Can you think of any additional studies you can make of the water mirage? The next time you see this mirage while driving, observe whether the image of another car ahead of you seems to be upright or inverted. From the diagram in Figure 50, which do you think it should be? How far ahead of you does the "water" appear to be? What does this apparent distance depend upon? When the mirage is seen, is the sun behind you or ahead of you? Does this make any difference? A systematic study of this phenomenon throughout the summer season will yield some interesting conclusions.

INVESTIGATION 24. (E)

Measuring Visibility

MATERIALS NEEDED:

1 tape measure, 50–100 foot length
1 flashlight bulb (3-volt), screw type
1 porcelain or plastic socket for screw-type flashlight bulb
2 flashlight dry cells
1 piece of white cardboard, 4 inches by 6 inches
1 small push-button switch
1 small variable resistor, 0–100 ohms
1 piece of plywood, ¾ inch by 12 inches by 12 inches
2 right-angle metal supports, 2 inches by 2 inches
8 wood screws, ¾ inch
2 small bolts, ⅛ inch by 1 inch
24 inches of copper wire, 20-gauge, insulated

One extremely practical aspect of atmospheric optics, especially for aviation, is the measurement of visibility. Even with instrumented landings, there is a moment just before touching down when the pilot must rely upon visual contact with the runway. At the landing speeds common for large aircraft—somewhere between 110 and 145 miles per hour—horizontal visibility during landings is a crucial factor for safety. Visibility is, of course, extremely important in automobile driving, as well. A sudden reduction in visibility due to snow, rain, fog, or dust can create dangerously uncertain driving conditions.

The International Meteorological Organization has defined daylight visibility as the "mean greatest distance toward the horizon that prominent objects such as mountains, buildings, trees, towers, etc., can be seen and identified by the unaided normal eye." Nighttime visibility is defined as the distance at which a light of 100 candlepower (approximately equal to a 150-watt frosted light bulb) becomes indistinguishable. In daylight or at night, any reduction in visibility results from the loss of light from object to observer by absorption, scattering, reflection, and veiling glare due to diffusion by particles in the line of sight.

To keep a systematic record of visibility conditions, first establish a sighting point surrounded by numerous landmarks at varying distances from the point of observation. These landmarks might be prominent trees, buildings, towers, power poles, industrial chimneys, or wooded areas, distant hills, and mountains. Select several landmarks in all directions from the observation point and be sure to pick some close as well as distant objects in each direction. Draw a map of the area, locating each object in its proper place. For distances up to 500 feet (150 meters), use a tape measure to measure the actual distance to the object, and record this information on your map. For distances greater than 500 feet, pace the distance in a straight line to the object, counting paces as you walk. To obtain the average length of your pace, measure the distance covered in a specific number of paces, such as 100, and divide the distance by that number.

Visibility is generally measured and reported on a scale ranging from 0 to 9. The table below gives this scale with the corresponding distances for daylight and nighttime visibility, according to the definition provided by the International Meteorological Organization.

Visibility

Scale Number	Daylight Visibility (meters)	Night Observations	
		Distance of Object	Distance at which a light of 100 c.p. becomes indistinguishable
0	Less than 50 m.	50 m. (.031 mi.)	100 m.
1	50-200 m.	200 m. (.125 mi.)	330 m.
2	200-500 m.	500 m. (.31 mi.)	740 m.
3	500-1000 m.	1 km. (.62 mi.)	1340 m.
4	1-2 km.	2 km. (1.25 mi.)	2.3 km.
5	2-4 km.	4 km. (2.5 mi.)	4.0 km.
6	4-10 km.	10 km. (6.25 mi.)	7.5 km.
7	10-20 km.	20 km. (12.5 mi.)	12 km.
8	20-50 km.		
9	Above 50 km.	At greater distances a 100 c.p. light is not suitable	

The normal relation between visibility and weather, with the corresponding scale numbers, is shown in the following table.

Normal Relation Between Weather and Visibility

Scale Number	*Daylight Visibility (meters)*	*Fog, Mist, or Haze*	*Snow*	*Drizzle*	*Rain*
0	Less than 50 m.	Dense	Very heavy	——	——
1	50–200 m.	Thick	Very heavy or heavy	——	——
2	200–500 m.	Medium	Heavy	——	Tropically heavy
3	500–1000 m.	Moderate	Moderate	Thick	Very heavy
4	1–2 km.	Mist	Light	Moderate	Heavy
5	2–4 km.	Slight mist or haze	Very light	Slight	Heavy
6	4–10 km.	Slight mist or haze	Very light	——	Moderate
7	10–20 km.	——	——	——	Light
8	20–50 km.	——	——	——	Very light
9	Above 50 km.	——	——	——	——

At the time of your observation, station yourself at the sighting point and scan the horizon in all directions for the selected landmarks. Record your estimates of visibility after noting which landmarks are visible and which are too distant to be seen under the prevailing conditions. Be sure to record the cause of reduced visibility, e.g., smoke, fog, haze, drizzle, snow, etc., if known. Your data sheet might look like this:

Time of Observation	*Estimated Visibility (Feet and Miles)* *N*	*E*	*S*	*W*
0800	6 mi. (haze)	6 mi. (haze)	4 mi. (haze)	¾ mi. (rain)

A simple device for estimating relative visibility at night can be constructed from a flashlight bulb (3 volt), a plastic or porcelain socket, two flashlight dry cells, a piece of white cardboard, a push-button switch, and a small variable resistor, which you can salvage from a discarded radio.

Using a piece of ¾-inch plywood as the mounting board, connect the light bulb, dry cells, switch, and variable resistor in series as shown in Figure 49. Fasten the cardboard to the plywood base in an upright position by using small metal right-angle supports. Attach the porcelain socket to the card-

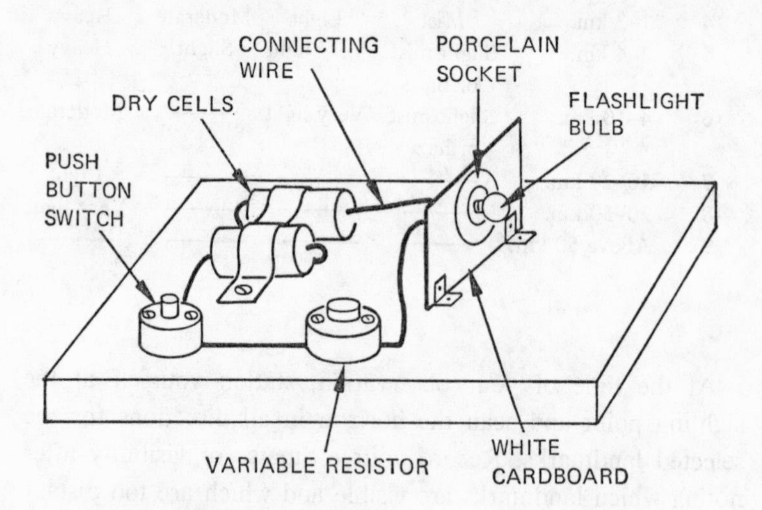

Fig. 49. Visibility Meter for Nighttime Measurements

board by small bolts. Attach the dry cells side by side by placing a thin metal strip over their tops and screwing the metal to the base. The push-button switch and variable resistor can be attached to the base by small screws.

To operate the apparatus, you will need to enlist the aid of an assistant. Set the resistor until the bulb burns at its brightest and have your assistant carry the apparatus away from you until the lighted bulb is just barely visible. While moving away from you, he should push the button switch at one-second intervals, keeping the bulb lighted for perhaps ½ second at a time.

When your assistant has gone such a distance from you that you fail to see the flashing light on 7 out of 10 trials, he has reached the limit of visibility, and the distance should be measured. For shorter distances, a measuring tape can be used. For greater distances, e.g., over 500 feet, use the pacing method described earlier.

The distances of relative visibility obtained by this method are based on the brightness of the light used in your apparatus. Suppose you used a brighter light. What would happen to your visibility distances? How can you compare your results with official visibilities reported by the U. S. Weather Bureau or an airline weather observer? Modify your apparatus to use a 100-candlepower bulb, and compare the results obtained with the standard distances reported in the table on page 184.

Some additional problems you might consider in further investigations of visibility are:

1. How does visibility vary with changing weather conditions, e.g., pressure, temperature, relative humidity, or frontal passages? Can you obtain any meaningful correlations?
2. How would you measure vertical visibility? To whom is

this information important? Can you devise an instrument to record vertical visibility directly through reduction of light? You might consider experimenting with light beams from a flashlight or projector.

3. Find out what are considered the safe minimum limits of visibility for commercial jet planes. CAVU means "Ceiling and Visibility Unlimited," that is, a ceiling over 9,751 feet and visibility over 10 miles. Will this 10-mile visibility still be adequate for SST's? (A supersonic transport plane traveling at 2,000 miles per hour will travel 10 miles in 18 seconds.)
4. How far away is a plane visible? How does background color and amount of daylight affect visibility?
5. What color should planes be for best visibility? What color should ships at sea be for best visibility? What color for least visibility (camouflage)?
6. What color pilot balloon (for wind observations) has the best visibility for various conditions?
7. What "colors" of light have more penetrating ability through haze and fog than visible light? How is wavelength related to penetrability?

Chapter 6

ATMOSPHERIC ELECTRICITY

The heavy, warm rainstorm, with its sharp cracks of thunder and bright flashes of lightning, is a familiar event, particularly in temperate latitudes. Almost all parts of the United States are visited by electrical storms at some time or another during any given year, although the frequency of these storms varies from less than 10 days per year on the West Coast to more than 70 days per year in Florida. This chapter will introduce several methods of studying electricity in the atmosphere, the phenomenon that makes its most dramatic manifestation in the giant spark of lightning that precedes a thunderclap. First, it will be helpful to review the theory of how these storms occur.

Two conditions frequently give rise to thunderstorms. We have mentioned earlier the process of convection, which occurs on hot summer days as the earth's surface gets heated and a warm layer of air near the ground starts to rise. The rising air cools and reaches its dew point when it forms cumulus clouds. If the air is moist, the clouds continue to grow and become cumulonimbus, or "thunder" clouds.

The second kind of thunderstorm forms as the sweeping passage of a cold or warm front displaces unstable air in its path. Thunderstorm cells of this type frequently occur in long lines parallel to the front and are more difficult to study by visual methods because of the confused nature of the sky and the great distances involved.

Two investigators from the University of Chicago, H. R. Byers and R. R. Braham, made a thorough study of thunderstorms during the late 1940s. By flying aircraft through the storms, they were able to learn that a typical storm has a 3-part life cycle: the cumulus, mature, and dissipating stages. During the cumulus stage, updrafts (rising air from below and to the sides) prevail, and the temperature of the thunderstorm is generally a few degrees warmer than that of the surrounding air. The cumulonimbus cloud may grow to a height of 30,000 to 40,000 feet. No rain falls during the cumulus stage, although precipitation particles are growing.

During the mature stage, updrafts and downdrafts occur in different parts of the storm cloud, and precipitation begins. The downdrafts bring cold air toward the ground and cause noticeable drops in temperature at the earth's surface. Lightning is usually most common during the mature stage, and hail may fall from the cloud.

The dissipating stage consists mainly of downdrafts and the decrease or complete cessation of rainfall. At this time the cloud begins to flatten out at the top, forming the characteristic anvil shape, and lightning subsides.

Although the lightning stroke accompanies a rainstorm, it is a manifestation of an electrical charge in the atmosphere that is present at all times, even on clear days and in all seasons. This charge consists of both positive and negative electricity, and changes in strength from summer to winter and in the course of a day.

The earth's surface is generally charged with negative electricity, and the air above it with positive electricity, although these charges are often reversed during thunderstorms. Because the earth and atmosphere are oppositely charged, there is an electric potential difference, or potential gradient, between points on the ground and points in the air that may amount to several hundred volts per meter of vertical distance over flat terrain. This is an average, however, and there may be large variations from place to place, especially in the vicinity of buildings, trees, or rough terrain. The potential gradient increases markedly just before an electrical storm, building to the huge spark between 2 regions of a cloud or between a cloud and the ground that we see as lightning.

Commercial electrometers are used at some weather stations or observatories to measure the potential gradient on a regular basis. Results of these measurements show that over the sea and level land areas, during fine weather, the average potential gradient is about 100 volts per meter. The annual variation is comparatively small in tropical regions but fairly large in temperate zones, where there is an increase to about 250 volts per meter during fall and winter, followed by a rapid decrease to about 100 volts per meter in the spring and summer. There is also a diurnal (daily) variation at all latitudes, with the highest potential gradient observed during daylight hours.

Measurements made with free balloons at very high elevations indicate that the potential gradient decreases with height. Estimates of the total potential difference between the earth and high atmosphere have been made as high as a million volts.

So far, attempts to find some relationship between potential gradient and various meteorological elements have not yielded conclusive evidence. However, the presence of various pollutants in the atmosphere, such as dust and smoke, seems to affect measurements greatly.

INVESTIGATION 25. **(M)**

Measuring Atmospheric Potential Gradient

MATERIALS NEEDED:

For apparatus setup: cigarette method

1 wooden dowel pole, 1½ inches by 6 feet
1 candle, ¾ inch by 6 inches
Several cigarettes
6 feet fine copper wire, uninsulated
1 metal grounding rod, ¾ inch by 3 feet

For apparatus setup: water-drop method

1 wooden block, ½ inch by 3 inches by 6 inches, drilled with ¾-inch hole
1 small can, single hole punched in bottom
1 candle, ¾-inch diameter
1 stiff wire, 8 inches long

For electroscope:

1 small wooden block, 1 inch by 3 inches by 3 inches
1 flat metal can, removable cover, such as typewriter ribbon case

1 thin metal strip, ½ inch by 6 inches
1 small insulating wax plug, 1 inch by 1 inch diameter, carved from paraffin block
1 paper clip formed as shown in Figure 52
1 tube of model airplane cement
1 pin or wire with eyelet, 1 inch long
2 cardboard strips 1 inch by 10 inches
1 small circular plastic or glass window, 2-inch diameter
Several small strips of aluminum foil, ⅛ inch by 1 inch
Solder, 1 small wood screw, ¾ inch

Potential gradient can be measured with fairly simple equipment. The apparatus shown in Figure 50 will give an accurate indication of the relative intensity of the potential gradient at different locations and times of the day. It consists of 3 essential components: an electrically conducting probe, which will take on the electric potential in the air; an electroscope, which will measure the potential gradient between the probe and the ground; and an insulator between the two.

The probe, or collector, is suspended from a pole at least 6 feet long and connected to the grounded electroscope by a fine wire. Two broom handles strapped end to end will provide sufficient height. The probe will reflect potential from the height of the pole. Should you wish to study the electrical potential at higher elevations, use a longer pole or run the connecting wire from the grounded electroscope to a probe suspended from, for example, a second-story window.

In order for the collector to assume the potential at the level of measurement and for the electroscope to indicate the difference in potential between that level and the ground, the

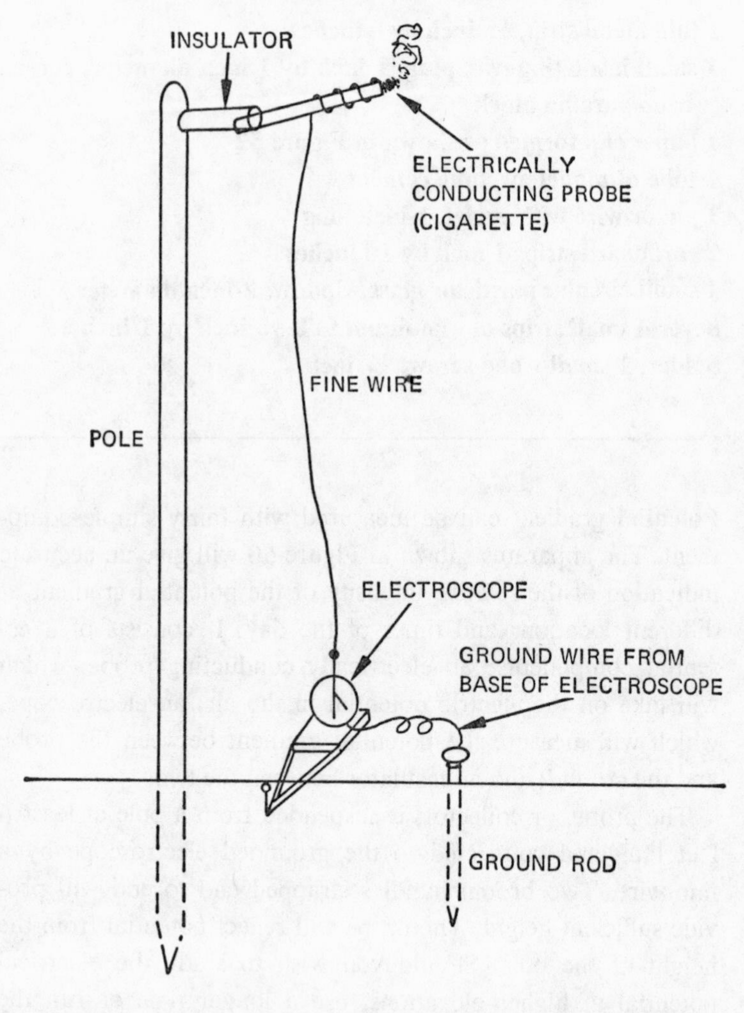

Fig. 50. Apparatus for Measuring Atmospheric Potential Gradient

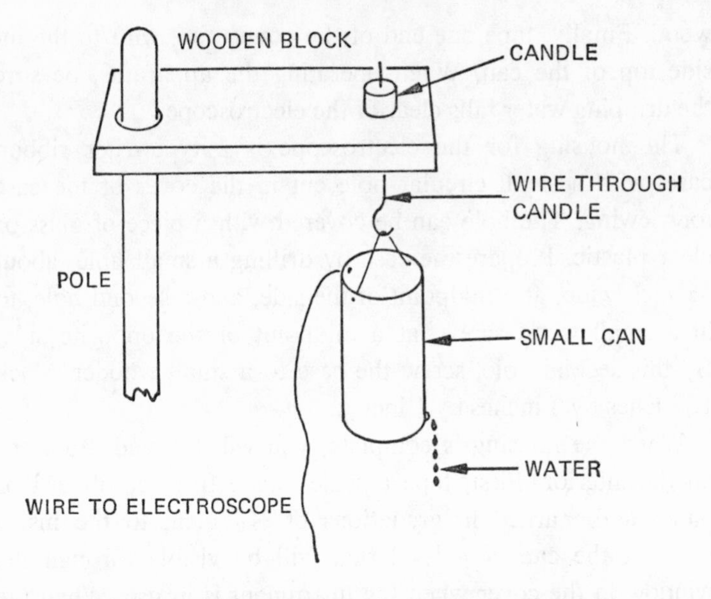

Fig. 51. Attachment of Can in Water-Drop Method

electric charge on the collector must be dispersed. Figures 50 and 51 show two kinds of collector that will serve this function. In the first, the gases and smoke of a lighted cigarette (a piece of burning rope will work as well) dissipate the charge. Tape the cigarette to an insulator—in this case, a candle about ¾ inch in diameter—and tape the candle to the top of the pole. Twist one end of a length of fine wire around the cigarette.

In Figure 51, the collector is a small can of water from which the water is permitted to drip slowly out a pinhole in the side, thus carrying away the charges. Again, the insulator is a sturdy candle. Hang the can from a wire extended through the candle, as shown, and push the candle through a hole in a small block of plywood. The block must be taped to the top of the pole, or wedge the pole through a second hole in the wood. Finally, tape one end of the connecting wire to the inside top of the can. When operating the apparatus, be sure the dripping water falls clear of the electroscope.

The housing for the electroscope is a typewriter ribbon case, with a small circular hole cut in the cover of the case for viewing. The hole can be covered with a piece of glass or clear plastic. Prepare the case by drilling a small hole, about ½ inch wide, at a midpoint of the side, and a second hole, to fit a small wood screw, at a midpoint of the opposite side. By this second hole, screw the case to a small wooden block (3 inches by 3 inches by 1 inch).

When the housing is complete, you will be ready to work on the interior. First, tape a scale, made from cardboard or paper and marked in gradations of 1/32 inch, to the inside back of the can at a level that will be visible through the window in the cover when the instrument is in use. When the scale is in place, glue a thin strip of metal, bent in the shape of an open-topped triangle, inside the housing, as indicated in Figure 52. The metal strip will make the electroscope more sensitive to electrical charge.

A paper clip serves as the rod and hanger for the instrument. Twist the paper clip into the shape indicated in the diagram, and from the lower portion hang a small aluminum foil leaf, which will function as the indicator. Make the leaf from foil peeled from a cigarette package or gum wrapper; household aluminum foil is too heavy. The leaf should be about ¾ inch long and 1/32 inch wide; it should be hung by a hook in one end of the foil so that it can swing freely in a plane parallel to the cover of the can. Force the straightened end of the paper clip through a small plug of wax fitted into the ½-inch hole at the top of the housing. This wax plug will serve to insulate the clip and aluminum foil assembly from

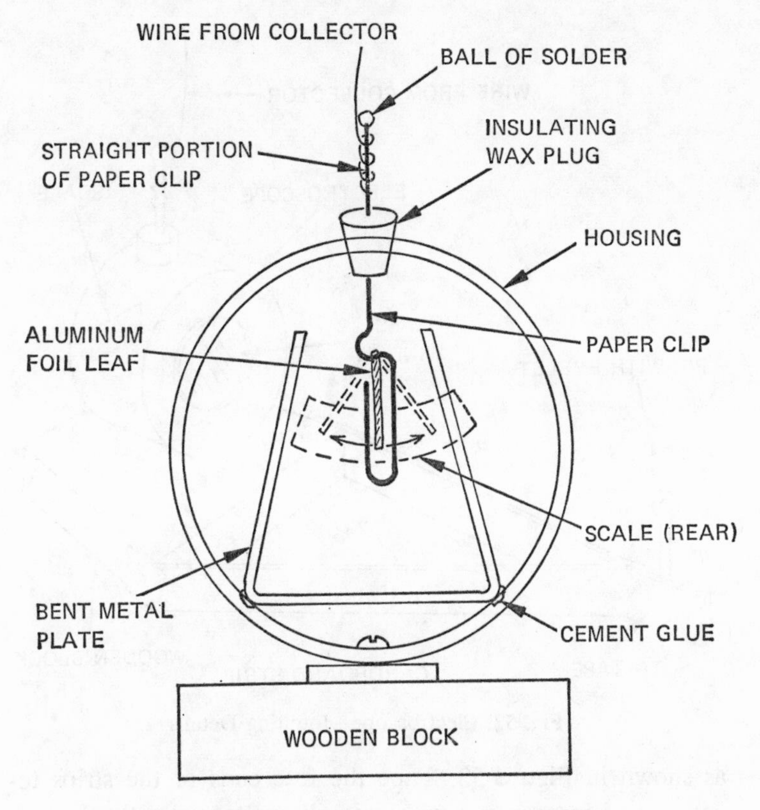

Fig. 52. Details of Electroscope Construction

other parts of the housing. To the top of the extended portion of the clip, affix a small ball of solder. Electric charge tends to dissipate from points or corners of charged objects, and the ball of solder will help minimize the leakage of charge from the electroscope. When you are ready to operate the instrument, wrap the wire from the probe around this extended portion of the paper clip.

Two further steps are necessary. To the front end of the base block, attach 2 cardboard strips (1 inch by 10 inches),

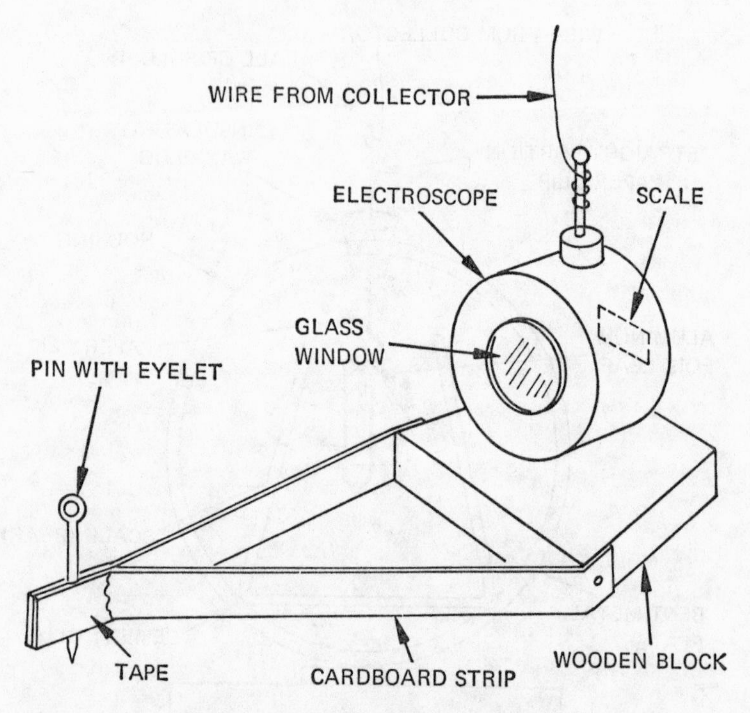

Fig. 53. Electroscope Mounting Details

as shown in Figure 53. Tape the free ends of the strips together and insert a pin with a small eyelet vertically through the tape. The eyelet will be used to position the eye for sighting the electroscope leaf. Finally, ground the electroscope by taping a wire to the housing and running the wire to a metal rod. The rod should be driven into a dampened portion of ground when you are ready to operate.

The electroscope can be charged by the following method: Rub a plastic comb on a piece of wool, and bring the charged comb a few inches from the loop of the electroscope. This should cause the aluminum foil to be deflected at an angle of about 45° from the paper-clip hanger. Keeping the comb in position, touch the loop with your finger. Remove your finger, then the comb. The electroscope should now have an induced positive charge, which will increase the sensitivity of the instrument. Small potential differences will deflect the foil either upward or downward.

To make your measurements, connect the wire from the collector to the grounded electroscope. Make sure that the thin wire hangs freely and does not come into contact with any object during the experiment. Light the cigarette, or fill the can with water, and raise the collector slowly until the electroscope leaf is deflected from its original position. Measure the height from the ground to the collector to the nearest inch. The reciprocal of this height is a relative measure of the potential gradient. (The greater the height, the smaller the potential gradient.) Or measure the amount of leaf deflection for different heights by the scale in the electroscope. This gives a direct measurement of the relative potential gradient at different heights.

What can you learn about the variations in potential gradient from place to place and from time to time? On a fair-weather day, measure the potential gradient at a number of locations around your home. Compare the readings you get in an open field, near a large tree, near your house, in a valley, on top of a hill, and near a body of water. Plot the readings on a map of the area surveyed. Take readings at different heights above the ground, such as at 1 meter, 2 meters, and 3 meters. Draw contour lines of equivalent potential gradient on a cross-sectional diagram showing various obstructions and varying topography. What seems to happen to the potential gradient near buildings?

Using the same location each time (preferably an open field), record the measurements of potential gradient each day for a month. Repeat the series of measurements during a different season, perhaps 6 months later. What changes do you observe? Try to correlate the readings of potential gradient with various meteorological elements, such as cloudiness, temperature, relative humidity, wind speed, and visibility.

Those readers who would like to build a more sensitive electrometer might consult an article entitled "How to Construct a Simple Electrometer," by Charles A. Laird, in the June 1949 issue of *Weatherwise*. Laird's apparatus consisted of a single vacuum tube, a D-200 microammeter, a dry cell, and a 22½-volt "B" battery, plus some varied sizes of resistors. This instrument can be used in rain as well as fair weather, and will indicate potential differences in the vicinity of thunderstorms, even when such storms are a number of miles from the location of the electrometer.

INVESTIGATION 26. (M)

Measuring Distances to Lightning Strokes

MATERIALS NEEDED:

1 protractor, at least 8-inch diameter
1 sighting stick (dowel), ¼ inch by 3 feet
Supply of polar-co-ordinate paper, 8½ inches by 11 inches

Many phenomena associated with electrical storms can be studied with reasonable safety while a storm is actually in progress. "Whistlers," for example—which we will get to later—are a direct result of electrical activity that can be detected many miles from the source. But one investigation that can be conducted within sight of a storm and that involves no more sophisticated equipment than a stop watch is the calculation of distance to lightning strokes. Since light travels almost a million times as fast as sound, the lightning flash is visible virtually at the moment it happens, while the thunderclap reaches the observer's ears perhaps seconds later. The time delay between the sight and the sound is a measurement of the distance between the observer and the storm.

During an electrical storm in your vicinity, observe a distant lightning flash, and time the interval to the first thunderclap. Multiply the number of seconds of time delay by 1,100 feet (the approximate speed of sound per second), and divide this number by 5,280 (the number of feet in a mile) to arrive at a value in miles.

For example, suppose the time delay noted on your stop watch is 6.5 seconds. Then the distance to the lightning stroke is:

$$6.5 \text{ seconds} \times \frac{1{,}100 \text{ feet per second}}{5{,}280 \text{ feet per mile}} = 1.35 \text{ miles}$$

For an approximate result, divide the number of seconds of time delay by five. In the example, the answer 1.4 is an accurate enough estimate, considering the probable error due to reaction time in starting and stopping the watch.

During an afternoon when a thunderstorm is in progress, make repeated observations of the distance to the storm, using the stop watch method. Can you assume the lightning always occurs in the same part of the storm? How large a ground area do you think a thunderstorm covers?

An analysis of the speed and direction in which a thunderstorm cell is moving can be made if you determine the direction to the storm cell from the point of observation every time you make a distance measurement. You will need a large protractor and sighting stick, as shown in Figure 54. The protractor can be mounted, in a plane parallel to the ground, on an eye-

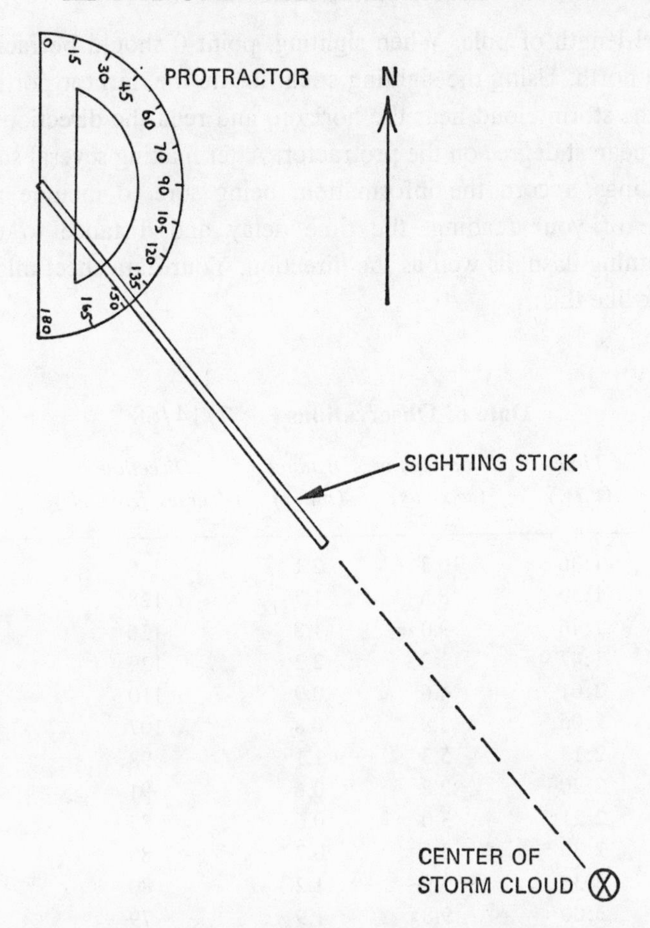

Fig. 54. Apparatus for Measuring Direction to Lightning Flash

level length of pole. When sighting, point 0 should be facing due north. Using the sighting stick, locate the darkest portion of the storm cloud near the horizon, and read the direction to the nearest degree on the protractor. After making several such readings, record the information, being sure to include the time of your reading, the time delay and distance to the lightning flash, as well as the direction. Your data sheet might look like this:

Date of Observations 7/14/68

Time (P.M.)	*Time Delay* (*seconds*)	*Distance* (*miles*)	*Direction* (*degrees from N.*)
1:36	10.3	2.1	140
1:39	8.6	1.7	128
1:46	9.0	1.8	126
1:47	11.2	2.2	120
2:01	4.6	0.9	110
2:06	3.8	0.8	107
2:13	5.3	1.1	98
2:20	2.8	0.6	91
2:21	3.0	0.6	85
2:30	3.6	0.7	80
2:38	5.9	1.2	80
3:00	9.8	1.9	79

By plotting these data on polar co-ordinate paper, as shown in Figure 55, you will be able to trace the approximate path of the thunderstorm.

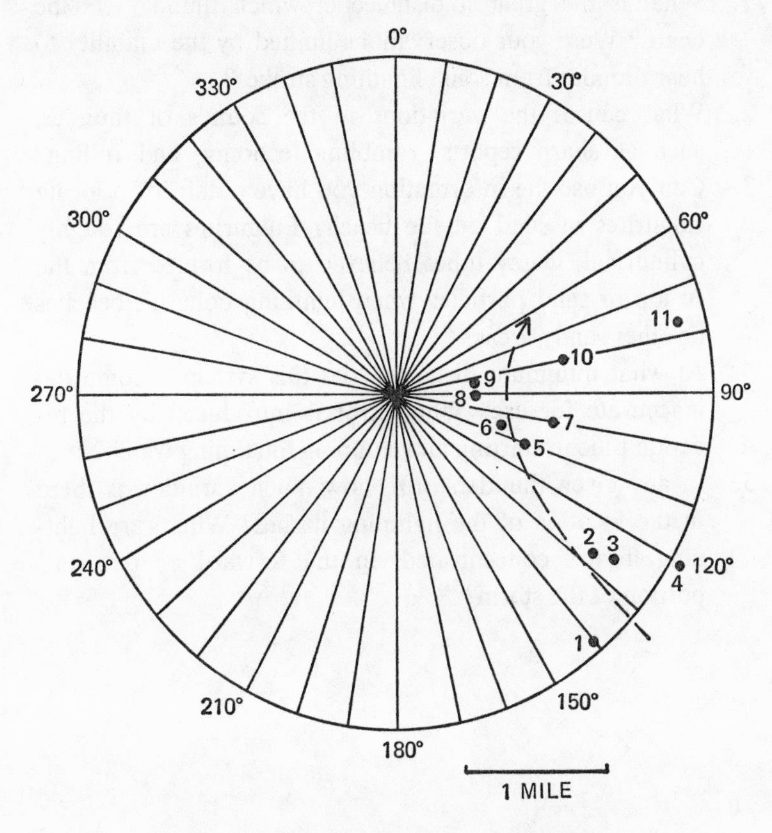

Fig. 55. Sample Plot of Path of a Thunderstorm

Here are some questions that you might consider during your investigation:

1. What is the greatest distance at which thunder can be heard? Were your observations limited by the inability to hear thunder from some lightning strokes?
2. What causes the variations in the sounds of thunder, such as sharp reports, rumbling, echoing, and rolling?
3. Can you use the information you have obtained to locate fulgurites in sand on the beach? Fulgurites are roughly cylindrical, glassy tubes believed to be formed from the fusion of sand particles when lightning bolts hit beaches or other sandy areas.
4. At what minimum distance does this system become too inaccurate for use? What error is introduced by the reaction time in starting and stopping the timing watch?
5. In any given thunderstorm, how much variation is there in the location of the lightning flashes? Where are lightning flashes concentrated, in the forward or rearward portion of the storm?

INVESTIGATION 27. (D)

Studying Whistlers

MATERIALS NEEDED:

1 hi-fi amplifier for receiver
1 antenna (fine copper wire, uninsulated, 200–300 feet long)

During World War I, Heinrich Barkhauser, a German physicist, heard strange whistling noises in some telephone wire-tapping equipment. At the time, the noises were attributed to some malfunction within the equipment itself, but further study during the 1920s and 1930s by Bell Telephone Laboratories and the Marconi Wireless Telegraph Company in England established that these "whistlers" had an atmospheric origin. It was discovered that "whistlers" are long, drawn-out whistling sounds that always follow a regular pattern, starting at a high pitch and lowering to about 500 cycles per second within a second or two after commencing. Associated with these whistling sounds are clicks of short duration that frequently precede the whistler by 1 or 2 seconds.

We now know that whistlers are the result of a sharp lightning report (heard as a click in the receiver), from which electromagnetic radiations have traveled an enormous distance, until, finally, the high and low frequencies have been spread apart due to differing speeds of travel. The consequent effect is the peculiar whistle, sounding from high to low pitch. It is believed that these radiations travel from one hemisphere to another along the earth's magnetic lines of force, traversing a distance of over 15,000 miles in a few seconds.

Peter Viemeister, in *The Lightning Book*, writes:

"The theory of the whistler was demonstrated in 1955 by Millett G. Morgan and G. McK. Allcock, using synchronized recording receivers in New Zealand and in the Aleutian Islands, which are on opposite ends of a line of magnetic force. On August 28, lightning struck New Zealand. Almost immediately the New Zealand receiver heard a sharp click. About one and one-half seconds later the Aleutian receiver heard a whistler. After about another one and one-half seconds the New Zealand receiver heard the whistler. Records covered a period of more than nine seconds, during which time each receiver heard the whistler at least three times, and each subsequent whistler was more drawn out and weaker than the one before."

Figure 56 shows the possible paths of radiation in the experiment Viemeister described. Originating in New Zealand, the whistler would have traveled to the Aleutian Islands, back to New Zealand, and once again to the Aleutians along the magnetic lines of force labeled 1, 2, and 3 in the diagram.

Whistlers can be studied on home-made equipment consisting of a simple receiver and a long antenna. The receiver can be an old hi-fi amplifier. The antenna, which must be several hundred feet long, can be a copper wire attached directly to the amplifier input terminals. No tuner is needed.

Because lightning storms are occurring continuously somewhere on the earth, whistlers can be observed during any

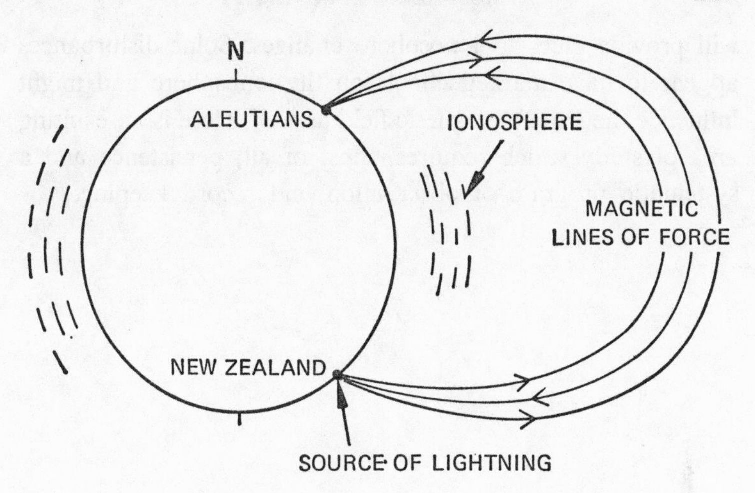

Fig. 56. Possible Paths of Radiation in a Whistler

season and at any time of day or night. Your receiver should be able to pick up clicks as well as whistlers. The clicks will be caused by lightning strokes in your vicinity—perhaps within a radius of 200 to 300 miles. The whistlers, on the other hand, may originate either in your own or the opposite hemisphere. What you hear may be the result of 1, 2, 3, or more traversals from one hemisphere to the other along magnetic field lines.

Keep a record of the frequency of whistlers and their associated clicks. Can you detect any cyclical nature to your record? Is there any difference between day and night recordings? Since the amplified sounds you hear are caused by lightning strokes, would you expect any daily variations?

Your observation of whistlers may also provide an inexpensive method for studying the ionosphere. Because the ionosphere transmits radiation signals, any changes in it will cause changes in the behavior of whistlers. Perhaps an analysis of the variations in pitch changes and the frequency of whistlers will provide clues to ionosphere changes. Solar disturbances appear to have marked effects on the ionosphere and might influence the earth's magnetic field as well. Here is an exciting area of study which requires, most of all, persistence and a systematic program of observation and record keeping.

Chapter 7

ATMOSPHERIC STRUCTURE AND COMPOSITION

Textbooks usually list the contents of the atmosphere as: nitrogen, 78.08 per cent; oxygen, 20.95 per cent; argon, 0.93 per cent; carbon dioxide, 0.03 per cent, and trace gases, 0.01 per cent. Although these percentages describe the composition of normal dry air, they do not include the water (gas, liquid, and solid), dusts, smoke, pollen grains, chemical pollutants, salts, fibers, and other ingredients also floating in the atmosphere. These latter materials play an important part in meteorology, particularly the three forms of water and the various particles that act as condensation nuclei. Several of the experiments in this chapter deal with methods of sampling the atmosphere and examining some of its lesser-known components.

The first investigation concerns temperature measurements of a seldom thought of area of the atmosphere—between the ground surface and a height of 72 inches, where most weather observations are taken. This region of "micrometeorology" can hold many surprises.

INVESTIGATION 28. (E)

Studying Micrometeorology

MATERIALS NEEDED:

1 wooden pole, 6 feet long
3 outdoor thermometers, Fahrenheit, 10° F to 110° F range
1 plywood board, 1 inch by 12 inches by 12 inches
1 3-inch wood screw
3 small wood screws to mount thermometers on the pole

Perhaps on a sunny afternoon you became aware that the air felt hotter when you stretched out on the grass or sat on a concrete curbstone. Or, early one morning, you may have sensed that the surface of the ground was much cooler than the air at head level. There may even have been frost on the ground.

Micrometeorology addresses itself to some difficult questions about this first few feet of atmosphere above the ground. How can the region be studied systematically? How widely do temperature and humidity vary during a 24-hour period? Are growing plants affected by these variations? How are small animals adapted to the conditions of the microclimate near the ground?

The average amount of energy received at the surface of the earth is about ½ langley per minute during daylight hours. (A langley equals 1 gram calorie per square centimeter.) Depending upon the absorption characteristics of the surface, the temperatures of various ground areas will differ markedly, even under uniform receipt of energy. Dark, rough surfaces, for example, tend to increase in temperature at a faster rate than light, smooth surfaces, because of their higher absorptivity. The air in contact with various surfaces likewise will have different temperatures due to the conduction of heat from the surface to the air immediately above it. At a distance of several inches above the surface, however, the air will be influenced by horizontal currents, and mixing will occur. Temperatures at this level may not be affected by the heating of the surface as much as the air nearer the ground.

An interesting experiment can be conducted to study the temperature variations in the region extending from the surface of the ground up to a height of 6 feet. You will need 3 inexpensive outdoor thermometers and a wooden pole 6 feet tall. Attach the thermometers to the pole at the base, at the 2½-foot height and at the 5-foot height, and screw the pole to a 1-foot-square wooden base. The apparatus will then stand vertically, and the thermometers will register the air temperatures at their respective heights.

Take temperature readings over different kinds of surfaces. By observing other atmospheric conditions, such as cloudiness, wind speed, or relative humidity, at the time of the readings, try to discover correlations between various elements of the weather.

An example of a similar experiment, conducted by Mr. Dale Rheel of Wisconsin, illustrates how such data can be collected and analyzed. In this instance, four different surface conditions were used: asphalt, gravel, grass, and concrete.

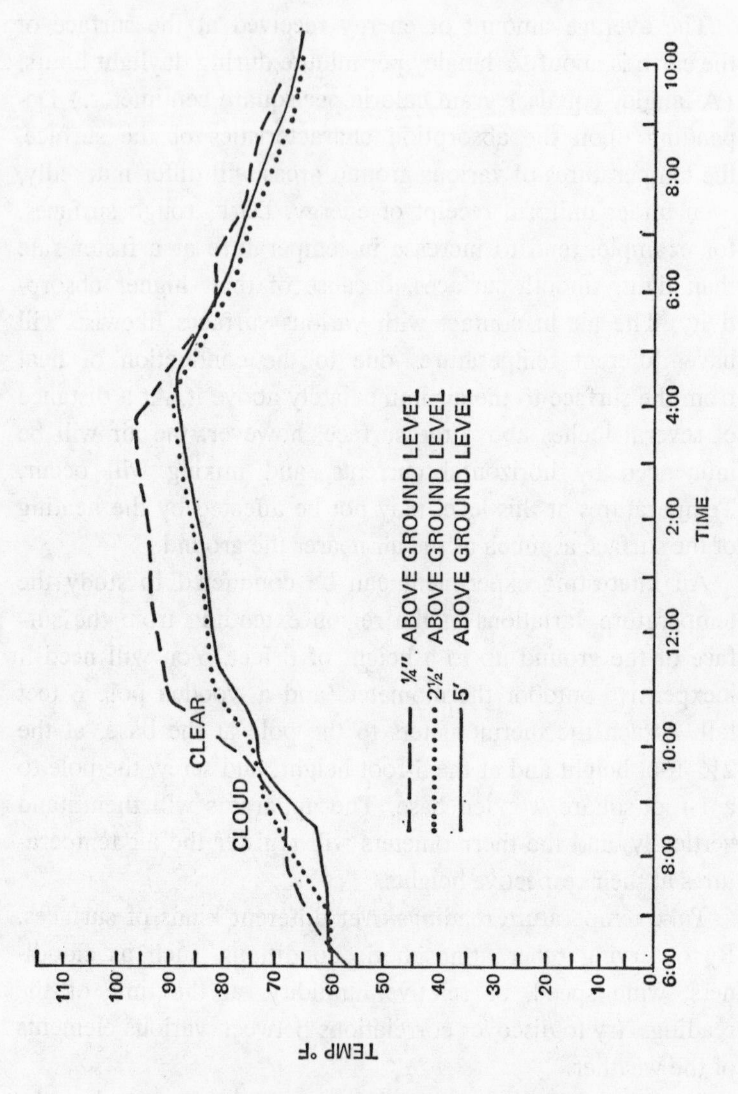

Fig. 57. Temperature Variation on Blacktop

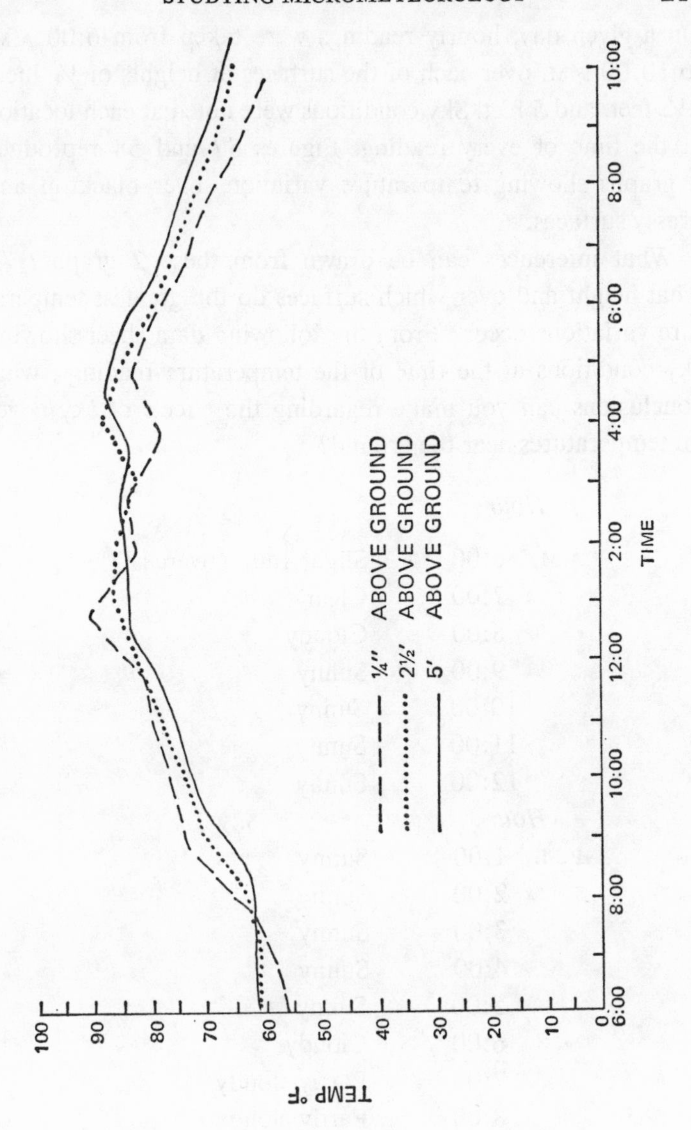

Fig. 58. Temperature Variation on Grass

On a given day, hourly readings were taken from 6:00 A.M. to 10:00 P.M. over each of the surfaces at heights of ¼ inch, 2½ feet, and 5 feet. Sky conditions were noted at each location at the time of every reading. Figures 57 and 58 reproduce 2 graphs showing temperature variations over blacktop and grassy surfaces.

What inferences can be drawn from these 2 graphs? At what height and over which surfaces do the greatest temperature variations occur? From the following data sheet showing sky conditions at the time of the temperature readings, what conclusions can you make regarding the effects of sky cover on temperatures near the ground?

	Hour	*Sky*
A.M.	6:00	Slight rain (overcast)
	7:00	Clear
	8:00	Cloudy
	9:00	Sunny
	10:00	Sunny
	11:00	Sunny
	12:00	Sunny
	Hour	*Sky*
P.M.	1:00	Sunny
	2:00	Sunny
	3:00	Sunny
	4:00	Sunny
	5:00	Sunny
	6:00	Cloudy
	7:00	Partly cloudy
	8:00	Partly cloudy
	9:00	Dark
	10:00	Dark

What modifications would you make to this experiment? How does height affect the readings? Is soil temperature the same as air temperature? How does air movement affect the readings? What is the relationship between relative humidity and temperature in the region studied? What is the effect on temperature of tall plants, such as corn or grains, in the area?

INVESTIGATION 29. (M)

Constructing an Airborne Rocket Thermograph

MATERIALS NEEDED:

Several model rockets (see Appendix C for address of the source)
1 small metallic thermometer
1 plastic fuel line (for model airplane engine) 2 inches long

---

I would like to describe a "fun" project involving model rockets that collect information about the temperatures aloft. With such information, you can calculate the temperature lapse rate (the rate of change of temperature with height) for the atmosphere on several successive days. If atmospheric conditions are markedly different during this period—if, for example, a cold front has passed over the area—you will be able to learn something about the depth of the cold air mass, temperature inversions aloft, and other characteristics of the atmosphere. With sufficient readings taken at different altitudes, you might be able to draw a temperature profile for the air mass at your location over a fixed period of time.

The rockets that perform most reliably for this experiment are obtainable from Estes Industries (see Appendix C for the address). Several models are available, including 1-, 2-, and 3-stage rockets. The experiment described here used a 3-stage rocket, and the average height obtained in 6 flights was 1,640 feet above the ground.

The "payload" in this rocket is a thermograph, which can measure the temperature at the highest point of the rocket flight. A small metallic thermometer, such as might be used

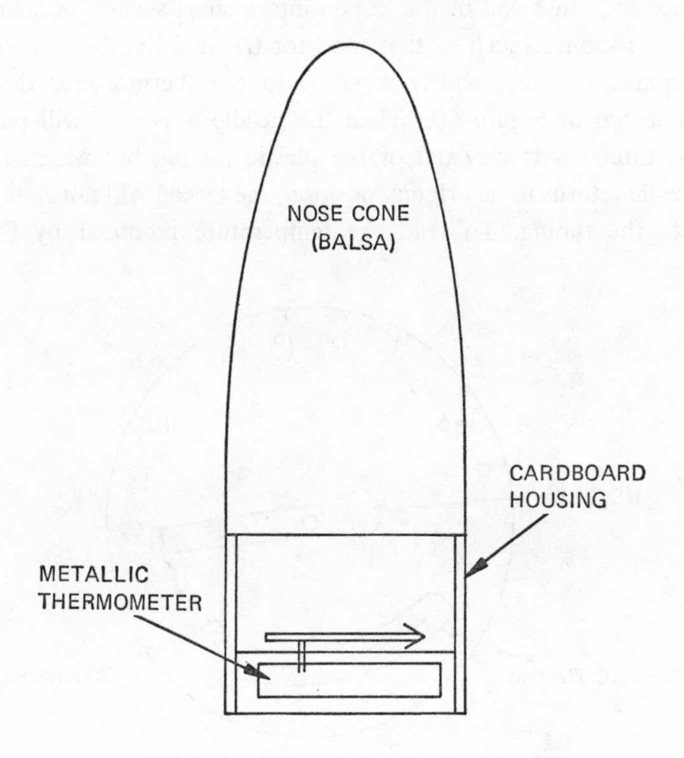

Fig. 59. Diagram of Payload Section

in oven cooking, works satisfactorily for this purpose. Trim the dial down so that its diameter is about 1¼ inches, but make sure the mechanism is still intact. The temperature scale can be reduced in size by painting the graduations closer to the center of the dial. Mount the thermometer horizontally in the "payload" section of the rocket (as shown in Figure 59), which will be retrieved by parachute after ejection at the top of the flight.

Some means of recording the temperature at the top of the flight will be necessary. One way is to attach a thin thread, about 2 inches long, to the tip of the thermometer needle. Feed the other end of the thread into a short section of small plastic tubing (such as that used for the fuel line in a model airplane engine), which is glued to the thermometer dial, as shown in Figure 60. When the needle moves, it will pull the thread part way out of the plastic tubing, but when the needle returns to its original position, the thread will not return into the tubing. To read the temperature recorded by the

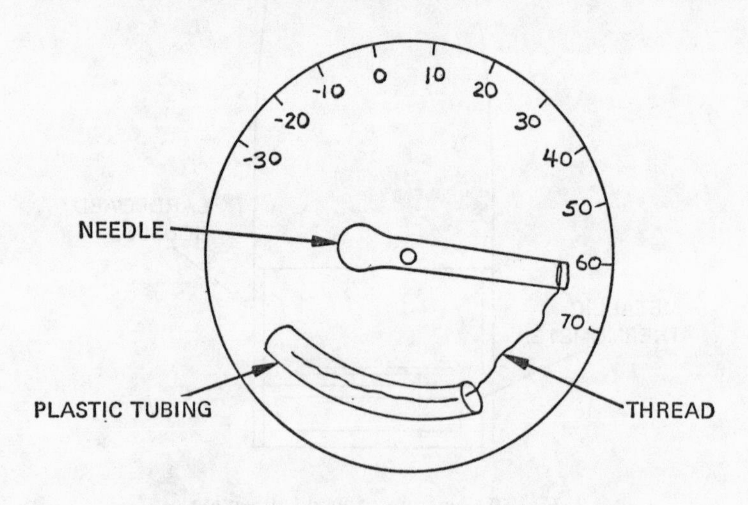

Fig. 60. Metallic Thermometer (top view—enlarged)

thermometer at the top of the flight, simply move the needle manually until the thread is taut, which will indicate how far the needle was deflected and, consequently, what the temperature was aloft.

The height of the rocket at the top of its flight can be determined by using a device for sighting the rocket at maximum altitude. The apparatus described earlier for measuring the heights of clouds is a suitable clinometer. Launch the rocket from the end of a base line approximately 1,000 feet long, and have an observer at the other end of the base line measure the angle of inclination, δ, between the sight line to the rocket and the horizon. The height can then be calculated by using the following equation:

$$H = (1000 \text{ feet}) (\tan \delta)$$

A series of 6 trial runs produced the following actual data:

Trial	*Date*	*Time*	*Temp. (ground)*	*Temp. (aloft)*	*Height*	*Comments*
1	2/26	3:13 P.M.	50° F	43° F	1,500 ft.	Partly cloudy Light wind
2	2/27	3:10 P.M.	20° F	10° F	1,250 ft.	
3	2/28	4:20 P.M.	20° F	10° F	1,760 ft.	West wind 5–10 mph
4	2/29	3:06 P.M.	46° F	43° F	1,440 ft.	
5	3/4	1:05 P.M.	44° F	41° F	1,680 ft.	Clear, no wind
6	3/4	2:00 P.M.	50° F	42° F	1,630 ft.	Clear, slight breeze

From these data one can see that the average height of the rocket was 1,640 feet, and the average distance in temperature beween the ground and the maximum height achieved by the rocket was 6.8° F. The average temperature lapse rate was 4.1° F per thousand feet. (The normal temperature lapse rate of the atmosphere is usually given as 3.5° F per 1,000 feet.)

By using rockets of varying power, capable of propelling the payload to a range of heights, you will obtain sufficient information to plot your data on a pseudoadiabatic chart, connecting the points showing temperature changes. You can prepare a simplified version of such a chart by using ordinary graph paper, and marking the abscissa (horizontal line) in units of temperature (Fahrenheit) and the ordinate (vertical line) in units of height (hundreds of feet). The commercial form of a pseudoadiabatic chart is more complicated, and values for pressure, potential temperature, equivalent potential temperature, and mixing ratio are plotted, as well as those for temperature and height.

INVESTIGATION 30. (M)

Constructing a Microbarometer

MATERIALS NEEDED:

1 Thermos bottle, 1-pint capacity
1 rubber or neoprene stopper to fit
1 thermometer, alcohol, 0° C to 50° C range
1 capillary tube, ⅛-inch bore by 50-inch length
1 small droplet of mercury
1 piece of ¼-inch glass tubing, 4 inches long, bent at right angle
1 short length of rubber tubing, ¼ inch by 4 inches
1 pinch clamp
1 plywood board, 6 inches by 50 inches
1 plywood board, ½ inch by 7 inches by 7 inches
Cleaning solution (100 milliliters concentrated sulfuric acid mixed with 4 milliliters saturated sodium dichromate)
12 small glass marbles
Silica gel (about 10 grams)

As the air is subjected to unequal heating in different latitudes or over varying land and water surfaces, local differences in atmospheric pressure develop. Pressures usually drop in response to heating or the addition of moisture, the former occurring because of the decrease in density as air is heated, and the latter because the density of water vapor is only about three fifths that of dry air.

Over the surface of the earth, there may be 50 to 100 relatively large areas of high or low pressure in existence at any one time. These are usually associated with cyclonic or anticyclonic motion of the air and are accompanied by fairly consistent weather variations. As a general rule, low pressures are indicative of cloudiness, high humidity, precipitation, and possible storms. On the other hand, high pressures usually signify clear skies; dry, cool air; and generally pleasant weather conditions.

An ordinary mercury barometer will accurately record daily fluctuations of atmospheric pressure. This is the measuring device used by weather services and was among the first instruments employed on a routine daily basis to predict atmospheric changes.

The usual long-range fluctuations in height of the mercury column in a barometer are less than 1 inch, although occasional severe storms, such as hurricanes or tornadoes, will cause greater fluctuations. Because the reader of an ordinary barometer must detect very small variations in the height of the mercury column, an accurate method of measuring minute height differences is necessary. The typical mercurial barometer can be read accurately to five thousandths of an inch through the use of movable vernier scales attached to the fixed scale of the barometer. The graduations on a vernier scale are only nine-tenths as long as those on the fixed scale.

Thus, precise measurements can be made for variations that are only one tenth as large as the smallest graduations on the fixed scale.

The barometer that I am about to describe, however, will register deflections of several inches rather than tiny fractions of an inch, thus usefully magnifying the reading of the barometric pressure change. Air replaces mercury as the working fluid. This form of barometer is based on the principle that a change in pressure produces a change in the volume of a gas when temperature is held constant. In the microbarometer, the temperature changes slowly because the air is contained in a vacuum flask. The change in volume due to a change in pressure is indicated by the motion of a mercury pellet located in a length of glass capillary tubing. When the atmospheric pressure increases, the pellet is forced downward; when the pressure decreases, the air in the vacuum flask expands, forcing the pellet upward. Because of the large ratio between the volume of the vacuum flask and the volume of the bore in the capillary tube, a small change in atmospheric pressure produces a relatively large change in the volume of air in the flask, and a correspondingly large change in the position of the pellet in the capillary tube.

The main piece of equipment in the microbarometer is the vacuum flask, which must contain the working fluid (air) under as constant a temperature as possible. A 2-pint Thermos bottle with a large rubber or neoprene stopper, made to fit tightly through the opening of the bottle, is suitable for this purpose. Three holes should be drilled through this stopper to accommodate the thermometer, capillary tube, and venting tube.

The capillary tube should be at least 50 inches long with a bore approximately ⅛ inch in diameter. When the apparatus is completed, a small droplet of mercury will be placed inside the capillary tube near the bottom. As the lower end of the tube is pushed through the stopper, the mercury droplet will be forced up the tube so that it is visible above the stopper. It is vital to keep the inside bore of this tube clean by removing grease or other foreign matter that will affect the smooth movement of the mercury pellet. Before inserting the mercury and operating the microbarometer, clean the capillary tube with a solution of 100 milliliters of concentrated sulfuric acid and 4 milliliters of saturated sodium dichromate ($Na_2Cr_2O_7$). Heat the solution before using it, and allow it to remain in the capillary tube about half an hour. Then rinse the tube with distilled water and allow it to dry.

The thermometer should be an alcohol thermometer, with a range of about 0° to 50° Centigrade.

The venting tube can be constructed from a piece of bent glass tubing, about 4 inches long, which has been fitted with a short length of rubber tubing and a pinch or screw clamp. After placing the 3 tubes through the stopper, make sure that they form an airtight seal.

Support the apparatus by a plywood frame, 6 inches by 50 inches, with an opening, about 3 inches by 8 inches, cut in the wood to accommodate the vacuum flask (see Figure 61). The frame then straddles the flask, while the capillary tube is secured to the surface of the wood by tape or a small metal bracket. The entire frame can be set on a plywood base, ½ inch by 7 inches by 7 inches, attached with small right-angle brackets.

To calibrate your microbarometer, and to obtain magnification that will permit measurement of normal daily fluctuations of pressure on the tube length of the instrument, it is necessary to regulate the volume of air inside the vacuum

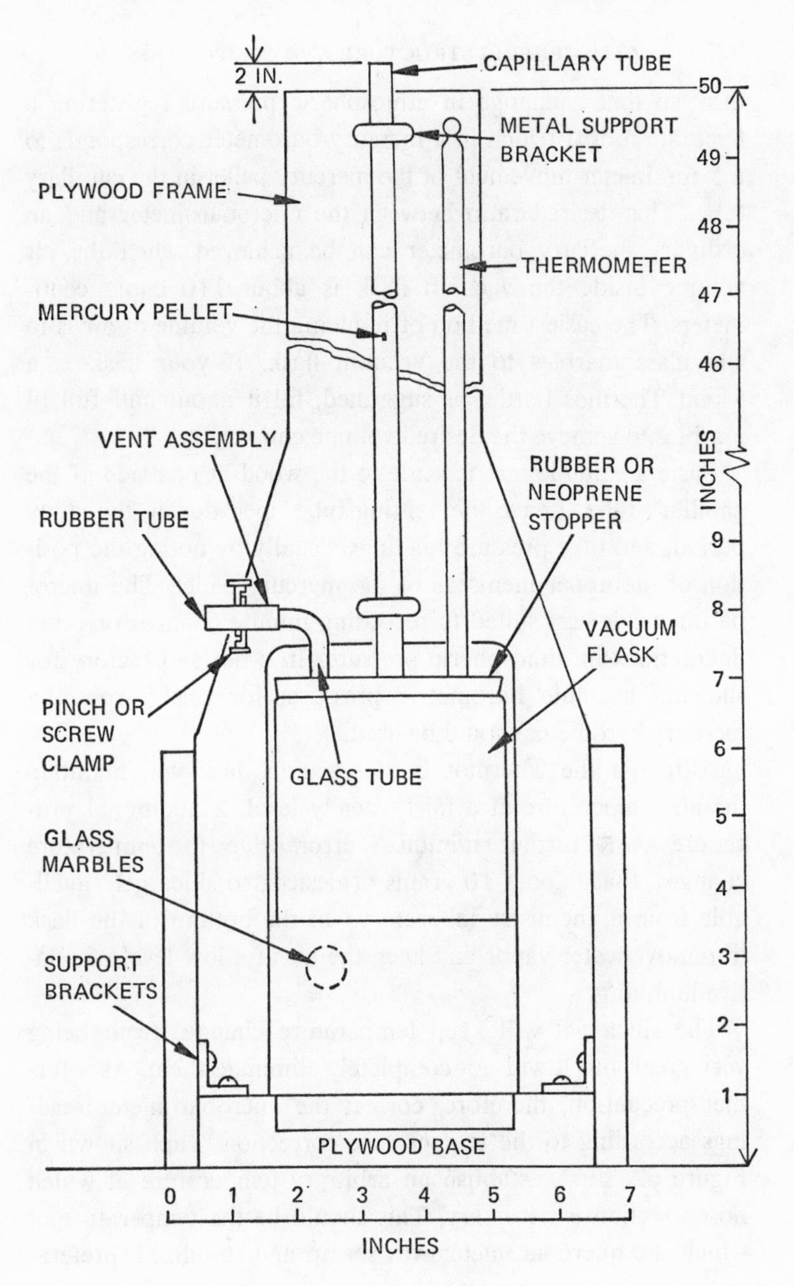

Fig. 61. Microbarometer Detail

flask so that a change in atmospheric pressure registering a fluctuation of 0.1 inch in a mercury barometer corresponds to a 5-centimeter movement of the mercury pellet in the capillary tube. This desired ratio between the microbarometer and an ordinary mercury barometer can be achieved when the air volume inside the vacuum flask is about 110 cubic centimeters. The easiest method of reducing the volume of air is to add glass marbles to the vacuum flask. If your flask is a 1-pint Thermos bottle, as suggested, fill it about half full of marbles to achieve the desired volume capacity.

Paste a suitable metric scale to the wood at one side of the capillary tube. Leave the venting tube open during the study period, and take pressure readings visually by noting the position of the upper meniscus of the mercury pellet. The microbarometer is best suited to recording minute changes or rapid fluctuations in atmospheric pressure. It is not satisfactory for showing absolute barometric pressure; for this purpose, a mercury barometer should be used.

Although the Thermos bottle-vacuum flask will maintain the air temperature at a fairly steady level, 2 additional procedures will further minimize errors due to temperature changes. Place about 10 grams of desiccated silica gel (available from a chemistry laboratory) in the bottom of the flask to remove water vapor and keep the air at a low level of relative humidity.

The silica gel will keep temperature changes from being very great, but it will not completely eliminate them. As a further precaution, therefore, correct the microbarometer readings according to the temperature correction graph shown in Figure 62. First, establish an arbitrary temperature at which no correction is necessary. This should be the temperature at which the microbarometer was set up and stabilized, prefera-

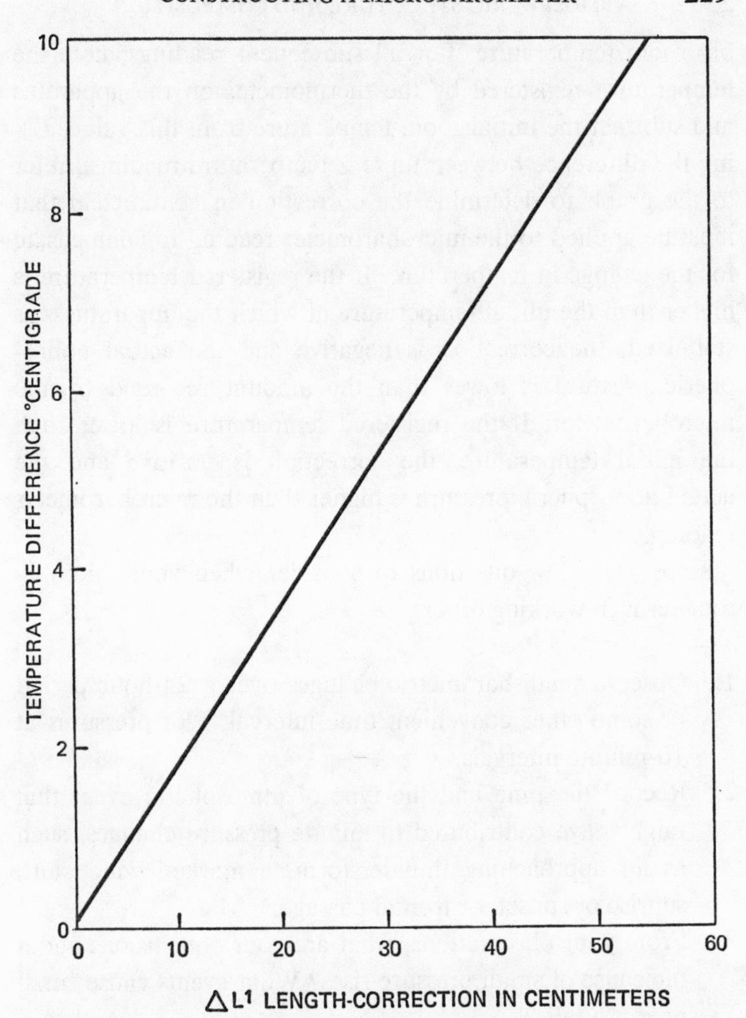

Fig. 62. Temperature Difference Centigrade

bly room temperature. For all subsequent readings, note the temperature registered by the thermometer on the apparatus and subtract the initial room temperature from this value. Using the difference between these 2 temperature readings, refer to the graph to determine the correction in centimeters that must be applied to the microbarometer reading to compensate for the change in temperature. If the registered temperature is higher than the initial temperature at which the apparatus was stabilized, the correction is negative and the actual atmospheric pressure is lower than the amount recorded by the microbarometer. If the registered temperature is lower than the initial temperature, the correction is positive and the actual atmospheric pressure is higher than the microbarometer reading.

Here are a few questions to consider when your microbarometer is in working order:

1. Observe small barometric changes over a 24-hour period or some other convenient time interval. Plot pressures at 10-minute intervals.
2. Record the time and the type of atmospheric event that might have contributed to minute pressure changes, such as an approaching thunderstorm, a marked wind shift, sunrise or sunset, or frontal passage.
3. From your observations, what are your conclusions about the cause of small pressure rises? What events cause small pressure falls?
4. How can you eliminate the effects of normal daily pressure changes from your data? Obtain records of diurnal (daily) pressure changes for your locality. Subtract the pressure reading from the diurnal graph for each point for which you have secured data. Plot a graph of these

differences. What is the purpose of doing this? Does your analysis of the probable causes of minor pressure fluctuations change with your new data?

5. Could minor earthquake shocks be recorded with a sensitive barometer? Obtain more information on the relationship between barometric pressure and earthquakes from the U. S. Weather Bureau.

INVESTIGATION 31. (D)

Measuring Salinity of the Atmosphere

MATERIALS NEEDED:

1 large glass container, 5-gallon capacity
1 2-hole rubber stopper to fit the container
1 Erlenmeyer flask, 500-milliliter capacity
1 2-hole rubber stopper to fit the flask
1 ring stand and support ring
Lengths of glass tubing, ¼-inch diameter, cut and bent as shown in Figure 63
1 siphon hose (rubber tubing, ¼-inch diameter, 6 feet long)
Small table to support apparatus
1 titration apparatus (2 burettes and burette stand)

Salt crystals in the atmosphere are among the most effective nuclei for condensation of water vapor. Because of their hygroscopic nature, condensation on these nuclei can begin at relative humidities far below 100 per cent, as was demonstrated in an earlier experiment. The investigation suggested here is a method for measuring the actual salt content of the atmosphere, using the apparatus illustrated in Figure 63.

The setup is designed to filter air through a container of distilled water, forming a solution of water and sodium chloride or any other soluble salt in the air, and simultaneously to measure the volume of air passing through the water. A 1-liter flask of distilled water is mounted on a ring stand. Inlet tube (A) admits air into the flask of water, and outlet tube (B) runs into a large airtight container with a capacity of about 20

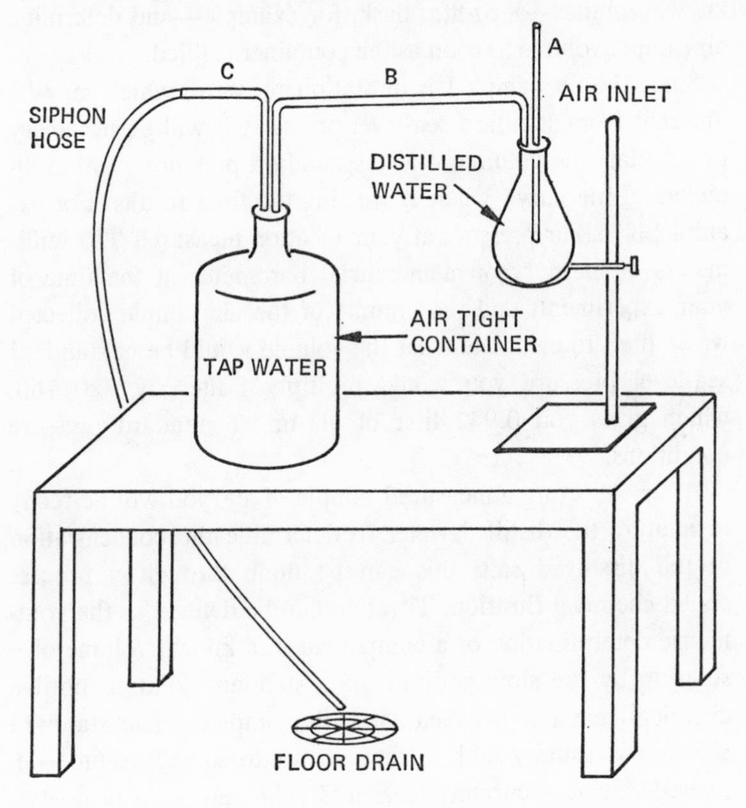

Fig. 63. Apparatus for Collecting Measured Samples of Air

liters (5 gallons), which has been filled with a known quantity of tap water. Water is drawn off from this container through a siphon hose (C) to produce a pumping action. The amount of air entering the flask of distilled water will be equivalent to the amount of water siphoned from the container. Thus, the volume of the air sample can be determined as soon as all the water has been siphoned from the airtight container. Or, instead of letting the siphoned water drain away, as illustrated, run the siphon hose into a container of known volume—a 10-liter flask, for example—and determine air sample volume as soon as the container is filled.

Since the air sampled is at station pressure, which may be different from standard sea-level pressure, it will be necessary to calculate the volume of air at standard pressure (760 millimeters of mercury) when expressing the final results. For example, if the air pressure at your location measured 720 millimeters of mercury on a mercurial barometer at the time of your experiment, and the volume of the air sample collected was 1 liter, to calculate what the volume would be at standard sea-level pressure you would multiply 1 liter by 720/760, which gives you 0.947 liter of air under standard pressure conditions.

After collecting a measured sample of air, you will be ready to analyze the distilled water to determine the concentration of the dissolved salt; this can be done through a process called chemical titration. Titration can be defined as the volumetric determination of a component in a known volume of a solution by the slow addition of a standard solution until a chemical reaction between the 2 is complete. The standard solution is usually added from a burette so the volume delivered can be accurately measured. The end point of a titration process is that stage at which there is an abrupt color change in an indicator. In this experiment, use as the indicator sodium chromate, which turns from yellow to pink or dull red when the chemical reaction is complete.

The normality of a solution is defined as the number of gram-equivalents per liter. Thus, a solution containing 1 gram-equivalent weight per liter of solution is a 1 normal solution. A gram-equivalent weight is defined as the number of grams of a solution equal to its equivalent weight. The equivalent weight can be obtained by dividing the formula weight by the number of reactive hydrogen atoms in the formula if it is an acid, or by the number of hydroxide ions if it is a base.

For example, the equivalent weight of sulfuric acid, H_2SO_4, is equal to:

$$\frac{\text{formula weight}}{\text{number of reactive hydrogen atoms}} = \frac{(2\times1)+(32)+(4\times16)}{2} = \frac{98.0}{2} = 49.0$$

and the equivalent weight of calcium hydroxide, $Ca(OH)_2$, is equal to:

$$\frac{\text{formula weight}}{\text{number of hydroxide ions}} = \frac{74.0}{2} = 37.0$$

If there is any confusion in your mind about the process of titration, it is advisable that you refer to a high school chemistry text and review the discussion on titration before beginning this experiment.

In the chemical titration process used here, the solution of sodium chloride in distilled water is titrated against a standard solution of silver nitrate, $AgNO_3$. The setup of the titration apparatus is illustrated in Figure 64. An accurate volume (approximately 100 milliliters) of salt water is taken from the distilled water flask and put into the beaker; a few drops of sodium chromate indicator are added to it. The burette is

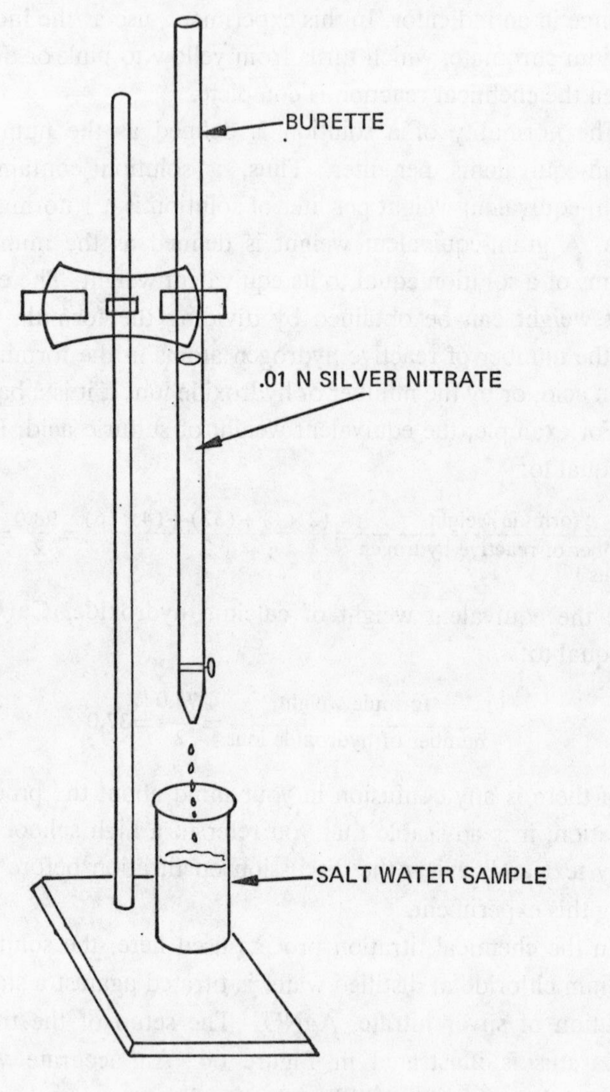

Fig. 64. Titration Apparatus

filled with a standard silver nitrate solution (0.01 N). You can prepare this yourself (see Appendix A for details) or obtain it from the local high school chemistry department.

After noting the initial reading on the burette, carefully add silver nitrate solution to the salt water solution drop by drop until the yellow sodium chromate shows a color change to pink or dull red. At this point, the chemical reaction is complete, and the final level of the silver nitrate in the burette should be noted.

A sample calculation is given below to illustrate the method by which the salt concentration of the distilled water can now be determined.

Volume of air passed through distilled water:	16 liters
Amount of distilled water:	1 liter
Normality of standard solution of $AgNO_3$:	0.01 N
Volume of $AgNO_3$ used in titration:	2.0 milliliters
Volume of water sample used in titration:	100.0 milliliters

At the end point of the titration, the product of the volume of silver nitrate times its normality equals the product of the volume of sodium chloride times its normality:

$$\text{Vol }(AgNO_3) \times N_{AgNO3} = \text{Vol }(NaCl) \times N_{NaCl}$$
$$(2.0 \text{ ml})\ (0.01) = (100 \text{ ml})\ N_{NaCl}$$
$$N_{NaCl} = \frac{(2.0)\ (0.01)}{100} = .0002 \text{ N}$$
$$N_{NaCl} = 2 \times 10^{-4} \text{ gram equivalents per liter of solution}$$

Therefore, the amount of sodium chloride per liter of solution equals:

$$(2 \times 10^{-4}) \times 58.5 = 117 \times 10^{-4} \text{ grams per liter}$$

(NOTE: 1 gram equivalent of NaCl=58.5 grams)

Because 16 liters of air were passed through 1 liter of distilled water to produce this concentration of sodium chloride in the solution, the concentration of sodium chloride in the air must have been:

$\frac{117\times10^{-4}}{16 \text{ liters}}$ = .00073 gram per liter of air,
or 0.73 gram per cubic meter of air,
or 3,222 tons of sodium chloride per cubic mile of air!

It now is perhaps a little easier to understand why the salinity of the atmosphere is a factor of such importance to meteorological processes. Hopefully, this investigation will lead you to some informed speculation on just how sodium chloride gets into the air in the first place. Your first thought, probably—and rightly—is the ocean.

Many of you have seen the pounding surf on an ocean coast. At the time the surf "breaks," a large amount of fine spray is thrown perhaps 5 or 10 feet into the air. Many of the very smallest water droplets evaporate completely before they return to the ground. Thus, along the seacoast, large numbers of tiny salt crystals are suspended in the air. These are caught up in large-scale air currents and carried high into the air, where they become part of the composition of the atmosphere. Because of this process of distribution, salt can be found in varying amounts in the atmosphere anywhere on the earth.

APPENDIX A

Derivation of equation $H=227\ (T_o-D_o)$, p. 43.

Let H=the height of the cloud bases,
T_o=the temperature at the ground surface,
D_o=the dew point at the ground surface,
T_h=the temperature at the cloud bases, and
D_h=the dew point at the cloud bases.

The lapse rate of temperature for a rising parcel of air is 5.5° F per 1,000 feet (adiabatic lapse rate). The lapse rate for the dew point is approximately 1.1° F per 1,000 feet. The following equations express the relationships between temperatures at the ground surface and at the cloud base, and similarly the relationships between the dew points at the ground surface and at the base of the clouds.

$$T_h=T_o-(5.5^\circ\ F/1{,}000\ \text{feet})\ H, \text{ and}$$
$$D_h=D_o-(1.1^\circ\ F/1{,}000\ \text{feet})\ H$$

Because H is the height at which condensation begins, T_h and D_h are the same temperature. Hence,

$$T_o-(5.5^\circ\ F/1{,}000\ \text{feet})\ H=D_o-(1.1^\circ\ F/1{,}000\ \text{feet})\ H$$
$$4.4\ H=1000\ (T_o-D_o)$$
$$H=227\ (T_o-D_o)$$

Instructions for making 0.01 normal silver nitrate solution:

Dissolve 1.7 grams of silver nitrate ($AgNO_3$) in 1,000 cubic centimeters (milliliters) of water. Place in dark brown bottle and label it.

Solution to Question 2, p. 131.

a. Calculate the Coriolis acceleration for the rotating cake pan with the following measurements:

Ω=30 revolutions per minute=½ revolution per second=180° per second
=radians per second=3.14 radians per second
V=10 centimeters per second (ejection velocity from the cans)
ϕ=90° latitude ($\sin \phi=1$)
d=diameter of cake pan (10 inches)

SOLUTION: Let a_{c_1} represent the Coriolis acceleration.

$$a_{c_1}=2\ \Omega\ V \sin \phi$$
$$=(2)(3.14/\text{sec.})(10\ \text{cm./sec.})(1)$$
$$=62.8\ \text{cm./sec}^2.$$

b. How would the Coriolis acceleration change by using a larger pan, say, 20 inches in diameter?

ANSWER: No change, since the equation for Coriolis acceleration ($a_c=2\ \Omega\ V \sin \phi$) does not consider diameter.

c. What would happen if the rate of rotation were increased to 45 rpm. (1½ times as fast as previously).

SOLUTION:

$$a_{c_2}=2\ \Omega\ V \sin \phi$$
$$=(2)(1\tfrac{1}{2}\times 3.14/\text{sec.})(10\ \text{cm./sec.})(1)$$
$$=94.2\ \text{cm./sec}^2.$$

APPENDIX B

Read down

TRIGONOMETRIC FUNCTIONS

Radians	Angle	Sin	Cos	Tan	Cot	Angle	Radians
0.000	0°	0.000	1.000	0.000	∞	90°	1.571
.018	1°	.018	1.000	.018	57.29	89°	1.553
.035	2°	.035	0.999	.035	28.64	88°	1.536
.052	3°	.052	.999	.052	19.08	87°	1.518
.070	4°	.070	.998	.070	14.30	86°	1.501
.087	5°	.087	.996	.088	11.43	85°	1.484
.105	6°	.105	.995	.105	9.514	84°	1.466
.122	7°	.122	.993	.123	8.144	83°	1.449
.140	8°	.139	.990	.141	7.115	82°	1.431
.157	9°	.156	.988	.158	6.314	81°	1.414
.175	10°	.174	.985	.176	5.671	80°	1.396
.192	11°	.191	.982	.194	5.145	79°	1.379
.209	12°	.208	.978	.213	4.705	78°	1.361
.227	13°	.225	.974	.231	4.331	77°	1.344
.244	14°	.242	.970	.249	4.011	76°	1.327
.262	15°	.259	.966	.268	3.732	75°	1.309
.279	16°	.276	.961	.287	3.487	74°	1.292
.297	17°	.292	.956	.306	3.271	73°	1.274
.314	18°	.309	.951	.325	3.078	72°	1.257
.332	19°	.326	.946	.344	2.904	71°	1.239
.349	20°	.342	.940	.364	2.747	70°	1.222
.367	21°	.358	.934	.384	2.605	69°	1.204
.384	22°	.375	.927	.404	2.475	68°	1.187
.401	23°	.391	.921	.425	2.356	67°	1.169
.419	24°	.407	.914	.445	2.246	66°	1.152
.436	25°	.423	.906	.466	2.145	65°	1.135
.454	26°	.438	.899	.488	2.050	64°	1.117
.471	27°	.454	.891	.510	1.963	63°	1.100
.489	28°	.470	.883	.532	1.881	62°	1.082
.506	29°	.485	.875	.554	1.804	61°	1.065
.524	30°	.500	.866	.577	1.732	60°	1.047
.541	31°	.515	.857	.601	1.664	59°	1.030
.559	32°	.530	.848	.625	1.600	58°	1.012
.576	33°	.545	.839	.649	1.540	57°	0.995
.593	34°	.559	.829	.675	1.483	56°	.977
.611	35°	.574	.819	.700	1.428	55°	.960
.628	36°	.588	.809	.727	1.376	54°	.943
.646	37°	.602	.799	.754	1.327	53°	.925
.663	38°	.616	.788	.781	1.280	52°	.908
.681	39°	.629	.777	.810	1.235	51°	.890
.698	40°	.643	.766	.839	1.192	50°	.873
.716	41°	.656	.755	.869	1.150	49°	.855
.733	42°	.669	.743	.900	1.111	48°	.838
.751	43°	.682	.731	.933	1.072	47°	.820
.768	44°	.695	.719	.966	1.036	46°	.803
.785	45°	.707	.707	1.000	1.000	45°	.785
Radians	Angle	Cos	Sin	Cot	Tan	Angle	Radians

Read up

Temperature of dew point in degrees Fahrenheit

[Pressure = 30.0 inches]

Air temperature t	Vapor pressure e	Depression of wet-bulb thermometer $(t - t')$														
		0.5	1.0	1.5	2.0	2.5	3.0	3.5	4.0	4.5	5.0	5.5	6.0	6.5	7.0	7.5
20	0.103	18	16	14	12	10	8	5	2	−2	−7	−13	−21	−37		
21	.108	19	18	16	14	12	9	7	3	±0	−4	−9	−16	−27	−60	
22	.113	20	19	17	15	13	11	8	5	+2	−2	−6	−12	−20	−36	
23	.118	21	20	18	16	14	12	10	7	4	±0	−4	−9	−16	−26	−57
24	.124	23	21	19	17	15	13	11	9	6	+2	−1	−6	−12	−20	−35
25	0.130	24	22	20	19	17	15	13	10	8	5	+1	−3	−8	−15	−25
26	.136	25	23	22	20	18	16	14	12	9	7	3	−1	−5	−11	−18
27	.143	26	24	23	21	19	18	16	13	11	8	5	+2	−2	−7	−14
28	.150	27	25	24	22	21	19	17	15	13	10	7	4	±0	−4	−9
29	.157	28	26	25	23	22	20	18	16	14	12	9	6	+3	−1	−5
30	0.164	29	27	26	25	23	21	20	18	16	14	11	8	5	+2	−2
31	.172	30	28	27	26	24	23	21	19	17	15	13	10	8	4	±0
32	.180	31	30	28	27	25	24	22	21	19	17	15	12	10	7	+3
33	.187	32	31	29	28	27	25	24	22	20	18	16	14	12	9	6
34	.195	33	32	30	29	28	26	25	23	22	20	18	16	13	11	8
35	0.203	34	33	31	30	29	28	26	25	23	21	19	17	15	13	10
36	.211	35	34	32	31	30	29	27	26	24	23	21	19	17	15	12
37	.219	36	35	33	32	31	30	28	27	26	24	22	21	19	17	14
38	.228	37	36	34	33	32	31	29	28	27	25	24	22	20	18	16
39	.237	38	37	35	34	33	32	31	29	28	27	25	23	22	20	18
40	0.247	39	38	37	35	34	33	32	30	29	28	26	25	23	21	20
41	.256	40	39	38	36	35	34	33	31	30	29	27	26	24	23	21
42	.266	41	40	39	38	36	35	34	33	31	30	29	27	26	24	23
43	.277	42	41	40	39	37	36	35	34	32	31	30	28	27	25	24
44	.287	43	42	41	40	38	37	36	35	34	32	31	30	28	27	25
45	0.298	44	43	42	41	40	38	37	36	35	34	32	31	30	28	27
46	.310	45	44	43	42	41	40	38	37	36	35	33	32	31	29	28
47	.322	46	45	44	43	42	41	40	38	37	36	35	33	32	31	29
48	.334	47	46	45	44	43	42	41	40	38	37	36	35	33	32	31
49	.347	48	47	46	45	44	43	42	41	40	38	37	36	34	33	32
50	0.360	49	48	47	46	45	44	43	42	41	40	38	37	36	34	33
51	.373	50	49	48	47	46	45	44	43	42	41	40	38	37	36	34
52	.387	51	50	49	48	47	46	45	44	43	42	41	40	38	37	36
53	.402	52	51	50	49	48	47	46	45	44	43	42	41	40	38	37
54	.417	53	52	51	50	49	48	47	46	45	44	43	42	41	40	38
55	0.432	54	53	52	51	50	50	49	48	47	45	44	43	42	41	40
56	.448	55	54	53	53	52	51	50	49	48	47	46	44	43	42	41
57	.465	56	55	54	54	53	52	51	50	49	48	47	46	45	43	42
58	.482	57	56	55	55	54	53	52	51	50	49	48	47	46	45	44
59	.499	58	57	56	56	55	54	53	52	51	50	49	48	47	46	45
60	0.517	59	58	57	57	56	55	54	53	52	51	50	49	48	47	46
61	.536	60	59	59	58	57	56	55	54	53	52	51	50	49	48	47
62	.555	61	60	60	59	58	57	56	55	54	53	53	52	51	50	48
63	.575	62	61	61	60	59	58	57	56	55	55	54	53	52	51	50
64	.595	63	62	62	61	60	59	58	57	57	56	55	54	53	52	51
65	0.616	64	63	63	62	61	60	59	59	58	57	56	55	54	53	52
66	.638	65	64	64	63	62	61	60	60	59	58	57	56	55	54	53
67	.661	66	65	65	64	63	62	62	61	60	59	58	57	56	55	54
68	.684	67	67	66	65	64	63	63	62	61	60	59	58	57	57	56
69	.707	68	68	67	66	65	64	64	63	62	61	60	59	59	58	57
70	0.732	69	69	68	67	66	65	65	64	63	62	61	61	60	59	58
71	.757	70	70	69	68	67	67	66	65	64	63	62	62	61	60	59
72	.783	71	71	70	69	68	68	67	66	65	64	64	63	62	61	60
73	.810	72	72	71	70	69	69	68	67	66	66	65	64	63	62	61
74	.838	73	73	72	71	70	70	69	68	67	67	66	65	64	63	62
75	0.866	74	74	73	72	71	71	70	69	68	68	67	66	65	64	64
76	.896	75	75	74	73	72	72	71	70	69	69	68	67	66	66	65
77	.926	76	76	75	74	73	73	72	71	71	70	69	68	67	67	66
78	.957	77	77	76	75	75	74	73	72	72	71	70	69	69	68	67
79	0.989	78	78	77	76	76	75	74	73	73	72	71	70	70	69	68
80	1.022	79	79	78	77	77	76	75	74	74	73	72	72	71	70	69

311278°—41—2

Temperature of dew point in degrees Fahrenheit

[Pressure = 30.0 inches]

Air temperature t	Vapor pressure e	Depression of wet-bulb thermometer $(t - t')$														
		8.0	8.5	9.0	9.5	10.0	10.5	11.0	11.5	12.0	12.5	13.0	13.5	14.0	14.5	15.0
25	0.130	−51														
26	.136	−32														
27	.143	−23	−45													
28	.150	−17	−29													
29	.157	−12	−20	−39												
30	0.164	−7	−14	−25	−57											
31	.172	−4	−10	−18	−31											
32	.180	−1	−6	−12	−21	−42										
33	.187	+2	−2	−7	−14	−26										
34	.195	5	+1	−3	−9	−17	−32									
35	0.203	7	4	±0	−5	−11	−20	−41								
36	.211	10	7	+3	−1	−6	−14	−25	−58							
37	.219	12	9	6	+2	−3	−8	−16	−29							
38	.228	14	11	8	5	+1	−4	−10	−19	−36						
39	.237	16	13	11	8	4	±0	−5	−12	−22	−47					
40	0.247	18	15	13	10	7	+3	−1	−6	−14	−26					
41	.256	19	17	15	12	10	6	+2	−2	−8	−16	−30				
42	.266	21	19	17	14	12	9	6	+2	−3	−9	−18	−36			
43	.277	22	20	19	16	14	11	9	5	+1	−4	−11	−21	−45		
44	.287	24	22	20	18	16	13	11	8	4	±0	−5	−12	−24	−60	
45	0.298	25	23	22	20	18	15	13	10	7	+4	−1	−6	−14	−27	
46	.310	27	25	23	21	20	17	15	13	10	7	+3	−2	−7	−16	−30
47	.322	28	26	25	23	21	19	17	15	12	9	6	+2	−3	−9	−17
48	.334	29	28	26	25	23	21	19	17	14	12	9	5	+1	−4	−10
49	.347	30	29	28	26	24	23	21	19	16	14	11	8	5	±0	−5
50	0.360	32	30	29	27	26	24	22	21	18	16	13	11	8	+4	±0
51	.373	33	32	30	29	27	26	24	22	20	18	16	13	10	7	+3
52	.387	34	33	32	30	29	27	26	24	22	20	18	16	13	10	7
53	.402	36	34	33	32	30	29	27	26	24	22	20	18	15	13	10
54	.417	37	36	34	33	32	30	29	27	25	24	22	20	18	15	12
55	0.432	38	37	36	34	33	32	30	29	27	25	24	22	20	17	15
56	.448	40	39	37	36	34	33	32	30	29	27	25	24	22	19	17
57	.465	41	40	39	37	36	34	33	32	30	29	27	25	24	21	19
58	.482	42	41	40	39	37	36	35	33	32	30	29	27	25	23	21
59	.499	44	43	41	40	39	37	36	35	33	32	30	29	27	25	23
60	0.517	45	44	43	41	40	39	38	36	35	33	32	30	29	27	25
61	.536	46	45	44	43	42	40	39	38	36	35	33	32	30	29	27
62	.555	47	46	45	44	43	42	40	39	38	36	35	33	32	30	29
63	.575	49	48	47	45	44	43	42	41	39	38	36	35	34	32	30
64	.595	50	49	48	47	46	44	43	42	41	39	38	37	35	34	32
65	0.616	51	50	49	48	47	46	45	43	42	41	40	38	37	35	34
66	.638	52	51	50	49	48	47	46	45	44	42	41	40	38	37	35
67	.661	53	53	52	50	49	48	47	46	45	44	43	41	40	38	37
68	.684	55	54	53	52	51	50	49	48	46	45	44	43	42	40	39
69	.707	56	55	54	53	52	51	50	49	48	46	45	44	43	42	40
70	0.732	57	56	55	54	53	52	51	50	49	48	47	46	44	43	42
71	.757	58	57	56	55	54	53	52	51	50	49	48	47	46	45	43
72	.783	59	58	58	57	56	55	54	53	52	51	50	48	47	46	45
73	.810	60	60	59	58	57	56	55	54	53	52	51	50	49	48	46
74	.838	62	61	60	59	58	57	56	55	54	53	52	51	50	49	48
75	0.866	63	62	61	60	59	58	57	56	55	55	54	52	51	50	49
76	.896	64	63	62	61	60	60	59	58	57	56	55	54	53	52	51
77	.926	65	64	63	62	62	61	60	59	58	57	56	55	54	53	52
78	.957	66	65	64	64	63	62	61	60	59	58	57	56	55	54	53
79	0.989	67	66	66	65	64	63	62	61	60	59	59	58	57	56	55
80	1.022	68	68	67	66	65	64	63	63	62	61	60	59	58	57	56

Relative humidity, per cent—Fahrenheit temperatures
[Pressure = 30.0 inches]

Air temperature t	Depression of wet-bulb thermometer ($t - t'$)																				
	0.5	1.0	1.5	2.0	2.5	3.0	3.5	4.0	4.5	5.0	5.5	6.0	6.5	7.0	7.5	8.0	8.5	9.0	9.5	10.0	10.5
20	92	85	77	70	62	55	48	40	33	26	19	12	5								
21	92	85	78	71	63	56	49	42	35	28	21	15	8	1							
22	93	86	78	71	65	58	51	44	37	31	24	17	11	4							
23	93	86	79	72	66	59	52	46	39	33	26	20	14	7	1						
24	93	87	80	73	67	60	54	47	41	35	29	22	16	10	4						
25	94	87	81	74	68	62	55	49	43	37	31	25	19	13	7	1					
26	94	87	81	75	69	63	57	51	45	39	33	27	21	16	10	4					
27	94	88	82	76	70	64	58	52	47	41	35	29	24	18	13	7	2				
28	94	88	82	76	71	65	59	54	48	43	37	32	26	21	15	10	5				
29	94	88	83	77	72	66	60	55	50	44	39	34	28	23	18	13	8	3			
30	94	89	83	78	73	67	62	56	51	46	41	36	31	26	21	16	11	6	1		
31	94	89	84	78	73	68	63	58	52	47	42	37	33	28	23	18	13	8	4		
32	95	89	84	79	74	69	64	59	54	49	44	39	35	30	25	20	16	11	7	2	
33	95	90	85	80	75	70	65	60	56	51	46	41	37	32	27	23	18	14	9	5	0
34	95	90	86	81	76	71	66	62	57	52	48	43	38	34	29	25	21	16	12	8	3
35	95	91	86	81	77	72	67	63	58	54	49	45	40	36	32	27	23	19	14	10	6
36	95	91	86	82	77	73	68	64	60	55	51	46	42	38	34	29	25	21	17	13	9
37	95	91	87	83	78	74	69	65	61	57	53	48	44	40	36	31	27	23	19	15	11
38	96	91	87	83	79	75	70	66	62	58	54	50	46	42	37	33	29	25	21	17	14
39	96	92	87	83	79	75	71	67	63	59	55	51	47	43	39	35	31	27	24	20	16
40	96	92	87	83	79	75	71	68	64	60	56	52	48	45	41	37	33	29	26	22	18
41	96	92	88	84	80	76	72	69	65	61	57	54	50	46	42	39	35	31	28	24	20
42	96	92	88	85	81	77	73	69	65	62	58	55	51	47	44	40	36	33	30	26	23
43	96	92	88	85	81	77	73	70	66	63	59	55	52	48	45	42	38	35	31	28	25
44	96	93	89	85	81	78	74	71	67	63	60	56	53	49	46	43	39	36	33	30	26
45	96	93	89	86	82	78	74	71	67	64	61	57	54	51	47	44	41	38	34	31	28
46	96	93	89	86	82	79	75	72	68	65	61	58	55	52	48	45	42	39	35	32	29
47	96	93	89	86	82	79	75	72	69	66	62	59	56	53	49	46	43	40	37	34	31
48	96	93	90	86	83	79	76	73	69	66	63	60	57	54	50	47	44	41	38	35	32
49	96	93	90	86	83	80	76	73	70	67	64	61	57	54	51	48	45	42	39	36	34
50	96	93	90	87	83	80	77	74	71	67	64	61	58	55	52	49	46	43	41	38	35
51	97	94	90	87	84	81	78	75	71	68	65	62	59	56	53	50	47	45	42	39	36
52	97	94	90	87	84	81	78	75	72	69	66	63	60	57	54	51	49	46	43	40	37
53	97	94	90	87	84	81	78	75	72	69	66	63	61	58	55	52	50	47	44	41	39
54	97	94	91	88	85	82	79	76	73	70	67	64	61	59	56	53	50	48	45	42	40
55	97	94	91	88	85	82	79	76	73	70	68	65	62	59	57	54	51	49	46	43	41
56	97	94	91	88	85	82	79	76	73	71	68	65	63	60	57	55	52	50	47	44	42
57	97	94	91	88	85	82	80	77	74	71	69	66	63	61	58	55	53	50	48	45	43
58	97	94	91	88	85	83	80	77	74	72	69	66	64	61	59	56	54	51	49	46	44
59	97	94	91	89	86	83	80	78	75	72	70	67	65	62	59	57	55	52	49	47	45
60	97	94	91	89	86	83	81	78	75	73	70	68	65	63	60	58	55	53	50	48	46
61	97	94	92	89	86	84	81	78	76	73	71	68	65	63	61	58	56	54	51	49	47
62	97	94	92	89	86	84	81	79	76	74	71	69	66	64	61	59	57	54	52	50	47
63	97	95	92	89	87	84	82	79	77	74	71	69	67	64	62	60	57	55	53	50	48
64	97	95	92	90	87	84	82	79	77	74	72	70	67	65	63	60	58	56	53	51	49
65	97	95	92	90	87	85	82	80	77	75	72	70	68	66	63	61	59	56	54	52	50
66	97	95	92	90	87	85	82	80	78	75	73	71	68	66	64	61	59	57	55	53	51
67	97	95	92	90	87	85	83	80	78	75	73	71	69	66	64	62	60	58	56	53	51
68	97	95	92	90	88	85	83	80	78	76	74	71	69	67	65	62	60	58	56	54	52
69	97	95	93	90	88	85	83	81	79	76	74	72	70	67	65	63	61	59	57	55	53
70	98	95	93	90	88	86	83	81	79	77	74	72	70	68	66	64	61	59	57	55	53
71	98	95	93	90	88	86	84	81	79	77	75	72	70	68	66	64	62	60	58	56	54
72	98	95	93	91	88	86	84	82	79	77	75	73	71	69	67	65	63	61	59	57	55
73	98	95	93	91	88	86	84	82	80	78	75	73	71	69	67	65	63	61	59	57	55
74	98	95	93	91	89	86	84	82	80	78	76	64	71	69	67	65	63	61	60	58	56
75	98	96	93	91	89	86	84	82	80	78	76	74	72	70	68	66	64	62	60	58	56
76	98	96	93	91	89	87	84	82	80	78	76	74	72	70	68	66	64	62	61	59	57
77	98	96	93	91	89	87	85	83	81	79	77	74	72	71	69	67	65	63	61	59	57
78	98	96	93	91	89	87	85	83	81	79	77	75	73	71	69	67	65	63	62	60	58
79	98	96	93	91	89	87	85	83	81	79	77	75	73	71	69	68	66	64	62	60	58
80	98	96	94	91	89	87	85	83	81	79	77	75	74	72	70	68	66	64	62	61	59

Relative humidity, percent—Fahrenheit temperatures

[Pressure = 30.0 inches]

Air temperature *t*	Depression of wet-bulb thermometer $(t - t')$																				
t	11.0	11.5	12.0	12.5	13.0	13.5	14.0	14.5	15.0	15.5	16.0	16.5	17.0	17.5	18.0	18.5	19.0	19.5	20.0	20.5	21.0
35	2																				
36	5	1																			
37	7	3																			
38	10	6	2																		
39	12	8	5	1																	
40	15	11	7	4	0																
41	17	13	10	6	3																
42	19	16	12	9	5	2															
43	21	18	14	11	8	4	1														
44	23	20	16	13	10	7	4	0													
45	25	22	18	15	12	9	6	3													
46	26	23	20	17	14	11	8	5	2												
47	28	25	22	19	16	13	10	7	5	2											
48	29	26	23	21	18	15	12	9	7	4	1										
49	31	28	25	22	19	17	14	11	9	6	3	1									
50	32	29	27	24	21	18	16	13	10	8	5	3	0								
51	34	31	28	26	23	20	17	15	12	9	7	4	2								
52	35	32	29	27	24	22	19	17	14	11	9	6	4	1							
53	36	33	31	28	26	23	20	18	16	13	10	8	6	3	1						
54	37	35	32	29	27	24	22	20	17	15	12	10	8	5	3	1					
55	38	36	33	31	28	26	23	21	19	16	14	12	9	7	5	2	0				
56	39	37	34	32	30	27	25	22	20	18	16	13	11	9	7	4	2				
57	40	38	35	33	31	28	26	24	22	19	17	15	13	10	8	6	4	2			
58	41	39	37	34	32	30	27	25	23	21	18	16	14	12	10	8	6	3	1		
59	42	40	38	35	33	31	29	26	24	22	20	18	16	13	11	9	7	5	3	1	
60	43	41	39	37	34	32	30	28	26	23	21	19	17	15	13	11	9	7	5	3	1
61	44	42	40	38	35	33	31	29	27	25	22	20	18	16	14	12	10	8	7	5	3
62	45	43	41	39	36	34	32	30	28	26	24	22	20	18	16	14	12	10	8	6	4
63	46	44	42	40	37	35	33	31	29	27	25	23	21	19	17	15	13	11	10	8	6
64	47	45	43	41	38	36	34	32	30	28	26	24	22	20	18	17	15	13	11	9	7
65	48	46	44	41	39	37	35	33	31	29	27	25	24	22	20	18	16	14	12	11	9
66	48	46	44	42	40	38	36	34	32	30	29	27	25	23	21	19	17	16	14	12	10
67	49	47	45	43	41	39	37	35	33	31	30	28	26	24	22	20	19	17	15	13	12
68	50	48	46	44	42	40	38	36	34	32	31	29	27	25	23	21	20	18	16	15	13
69	51	49	47	45	43	41	39	37	35	33	32	30	28	26	24	23	21	19	18	16	14
70	51	49	48	46	44	42	40	38	36	34	33	31	29	27	25	24	22	20	19	17	15
71	52	50	48	46	45	43	41	39	37	35	33	32	30	28	27	25	23	22	20	18	17
72	53	51	49	47	45	43	42	40	38	36	34	33	31	29	28	26	24	23	21	19	18
73	53	51	50	48	46	44	42	40	39	37	35	34	32	30	29	27	25	24	22	20	19
74	54	52	50	48	47	45	43	41	39	38	36	34	33	31	29	28	26	25	23	21	20
75	54	53	51	49	47	45	44	42	40	39	37	35	34	32	30	29	27	26	24	23	21
76	55	53	51	50	48	46	44	43	41	39	38	36	34	33	31	30	28	27	25	24	22
77	56	54	52	50	48	47	45	43	42	40	39	37	35	34	32	31	29	28	26	25	23
78	56	54	53	51	49	47	46	44	43	41	39	38	36	34	33	31	30	28	27	25	24
79	57	55	53	51	50	48	46	45	43	42	40	38	37	35	34	32	31	29	28	26	25
80	57	55	54	52	50	49	47	45	44	42	41	39	38	36	35	33	32	30	29	27	26

t	$(t - t')$														
t	21.5	22.0	22.5	23.0	23.5	24.0	24.5	25.0	25.5	26.0	26.5	27.0	27.5	28.0	28.5
61	1														
62	2	1													
63	4	2	0												
64	6	4	2	0											
65	7	5	4	2	0										
66	9	7	5	3	2	0									
67	10	8	7	5	3	2									
68	11	10	8	6	5	3	1								
69	13	11	9	8	6	5	3	1							
70	14	12	11	9	8	6	4	3	1						
71	15	13	12	10	9	7	6	4	3	1					
72	16	15	13	12	10	9	7	6	4	3	1				
73	17	16	14	13	11	10	8	7	5	4	3	1			
74	18	17	15	14	13	11	10	8	7	5	4	3	1		
75	20	18	17	15	14	12	11	9	8	7	5	4	3	1	
76	21	19	18	16	15	13	12	11	9	8	6	5	4	3	1
77	22	20	19	17	16	14	13	12	10	9	8	6	5	4	3
78	23	21	20	18	17	16	14	13	11	10	9	8	6	5	4
79	23	22	21	19	18	17	15	14	13	11	10	9	7	6	5
80	24	23	22	20	19	18	16	15	14	12	11	10	9	7	6

APPENDIX C

Source Materials (Equipment and Charts)

CHAPTER 2

INVESTIGATION 1

1. You can obtain mercury from:
   Sargent-Welch Scientific Company
   7300 North Linder Avenue
   Skokie, Illinois 60076

INVESTIGATION 2

2. A chart of the standard cloud types is available from:
   U. S. Weather Bureau
   Environmental Science Services Administration
   Washington, D.C. 20230

INVESTIGATION 3

3. You can obtain a plastic aquarium from:
   Ward's Natural Science Establishment, Inc.
   P. O. Box 1712
   Rochester, New York 14603

4. A pycnometer (specific gravity bottle) can be obtained from:
   Sargent-Welch Scientific Company
   7300 North Linder Avenue
   Skokie, Illinois 60076

INVESTIGATION 4

5. Psychrometric tables (Bulletin No. 235) may be ordered from:
   Psychrometric Tables
   Supt. of Documents
   U. S. Government Printing Office
   Washington, D.C. 20402

CHAPTER 3

INVESTIGATION 7

1. Fahrenheit thermometers can be obtained from:
   Sargent-Welch Scientific Company
   7300 North Linder Avenue
   Skokie, Illinois 60076

INVESTIGATION 9

2. Silver wire for the dew cell can be obtained from:
   Sargent-Welch Scientific Company
   7300 North Linder Avenue
   Skokie, Illinois 60076
3. A filament transformer can be obtained from:
   LaPine Scientific Company
   6001 South Knox Avenue
   Chicago, Illinois 60629
4. Further information on the construction of a dew cell may be obtained by writing to:
   The Foxboro Company
   Foxboro, Massachusetts 02035
   and
   Honeywell, Inc.
   2954 Fourth Avenue
   Minneapolis, Minnesota 55408

5. For information on lithium chloride humidity measuring devices, write to:

   American Instrument Company, Inc.
   Silver Springs, Maryland 20907

INVESTIGATION 10

6. An analytical balance can be obtained from:

   Van Waters and Rogers, Inc.
   P. O. Box 1050
   Rochester, New York 14603

INVESTIGATION 11

7. Alloys for constructing a thermopile are available from:

   The Sigmund-Cohn Corporation
   Mount Vernon, New York 10551

INVESTIGATION 14

8. A microscope can be obtained from:

   Sargent-Welch Scientific Company
   7300 North Linder Avenue
   Skokie, Illinois 60076

CHAPTER 4

INVESTIGATION 15

1. "Crystal Violet" can be obtained from:

   E. H. Sargent
   4647 West Foster Avenue
   Chicago, Illinois 60630

INVESTIGATION 17

2. A heater coil can be obtained in any well-stocked hardware store.

INVESTIGATION 18

3. A commercial wind meter is available from:
   Sargent-Welch Scientific Company
   7300 North Linder Avenue
   Skokie, Illinois 60076

INVESTIGATION 20

4. Small pressure tanks of helium can be purchased from:
   Sargent-Welch Scientific Company
   7300 North Linder Avenue
   Skokie, Illinois 60076
   and
   Central Scientific Company
   2600 South Kostner Road
   Chicago, Illinois 60613
5. Binoculars can be purchased in a department store or camera shop.

INVESTIGATION 21

6. Angle iron (wire) can be purchased in a hardware store or aquarium shop.

CHAPTER 5

INVESTIGATION 22

1. A glass aquarium can be purchased from:
   Sargent-Welch Scientific Company
   7300 North Linder Avenue
   Skokie, Illinois 60076

CHAPTER 7

INVESTIGATION 29

1. Rockets for constructing the airborne rocket thermograph can be purchased from:

   Estes Industries
   Box 227
   Penrose, Colorado 81240

2. A metallic thermometer can be purchased from:

   Sargent-Welch Scientific Company
   7300 North Linder Avenue
   Skokie, Illinois 60076

   or

   from a hardware store

3. A commercial pseudoadiabatic chart (Chart III Pseudoadiabatic Chart, 10 cents each) can be purchased from:

   Larue Printing Company
   Kansas City, Missouri 64127

INVESTIGATION 31

4. Burettes may be obtained from:

   Sargent-Welch Scientific Company
   7300 North Linder Avenue
   Skokie, Illinois 60076

   or

   Van Waters and Rogers, Inc.
   P. O. Box 1050
   Rochester, New York 14603

BIBLIOGRAPHY

Books

Battan, Louis J. *The Nature of Violent Storms.* Doubleday & Company, Inc., Garden City, New York, 1961.

A well-written book for junior and senior high students giving authoritative information on the awesome meteorological storms of nature. Excellent reference.

Bellaire, Frank R. and Stohrer, Albert W. *The Development of Laboratory and Demonstration Equipment for Meterorological Instruction.* The University of Michigan, Ann Arbor, Michigan, 1963.

This is a publication produced under an NSF grant in 1963 that describes the development and construction of some simple demonstrations simulating atmospheric phenomena. The level is considered to range from high school through junior college.

Byers, Horace R. and Braham, Roscoe R., Jr. *The Thunderstorm.* Supt. of Documents, U. S. Government Printing Office, Washington, D.C., 1949.

Report of research on thunderstorms carried out by the authors in 1948–49. Excellent source of authentic information on thunderstorm stages of development and electrical effects.

Greenstone, Arthur W.; Sutman, Frank X.; and Hollingworth, Leland G. *Concepts in Chemistry*. Harcourt, Brace, and World, Inc., New York, 1966.

A standard high school chemistry textbook. Contains information on normal and molar solutions, titration, and chemical equilibrium.

Instructions for Making Pilot Balloon Observations. W. B. 1278 Circular O, Aerological Division, U. S. Government Printing Office, Washington, D.C.

Gives details of conducting pilot balloon observations for determinations of wind speed and directions aloft. Describes standard equipment used by Weather Bureau.

Johnson, John C. *Physical Meteorology*. The Technology Press of M.I.T., 1954.

An advanced level book on physical bases of meteorology for college students.

Lane, Frank W. *The Elements Rage*. Chilton Company Publishers, Philadelphia, 1965.

This is a book about extreme natural violence. It includes storms, volcanoes, earthquakes, floods, and other catastrophic events. Stories are interestingly told with eyewitness accounts liberally used to document the events described.

Malan, D. J. *Physics of Lightning*. The English Universities Press, Ltd., London, 1963.

Authoritative book on electrical phenomena in the atmosphere. Useful as a reference source.

Neuberger, Hans and Nicholas, George. *Manual of Lecture Demonstrations, Laboratory Experiments, and Observational Equipment for Teaching Elementary Meteorology in Schools and Colleges*. Pennsylvania State University, University Park, Pennsylvania, 1962, p. 127.

A resourceful collection of simple yet clever devices to be constructed from inexpensive materials. Many diagrams and construction details given.

Petterssen, Sverre. *Introduction to Meteorology.* McGraw-Hill, New York, 1969.

A standard introductory textbook on meteorology, now in its third edition. A good reference for high school or first-year college students.

Reiter, Elmar R. *Jet Streams.* Doubleday & Company, Inc., Garden City, New York, 1967.

An authoritative book written for the high school student or interested lay reader. Comprises a comprehensive account of an interesting meteorological phenomenon of current interest.

Viemeister, Peter E. *The Lightning Book.* Doubleday & Company, Inc., Garden City, New York, 1961, pp. 271–74.

This book explores many interesting aspects of lightning and other forms of atmospheric electricity. Written in interesting style and is suitable for high school reading.

Periodicals

Acheson, D. T. "Some Limitations and Errors Inherent in the Use of the Dew Cell for Measurement of Atmospheric Dew Points." *Monthly Weather Review,* Vol. 91, No. 5, May, 1963, pp. 183–90.

Conover, J. H. "Tests and Adaptation of the Foxboro Dew-Point Recorder for Weather Observatory Use." *Bulletin American Meteorological Society,* Vol. 31, No. 1, January 1950, pp. 13–22.

Cox, Robert E. "Cloud Photography." *Weatherwise,* Vol. 3, No. 3, June 1950, 51–55.

Fultz, David. "A Survey of Certain Thermally and Mechanically Driven Systems of Meteorological Interest, Fluid Models in Geophysics." *Proceedings of First Symposium on the Use of Models in Geophysical Fluid Dynamics,* American Meteorological Society, 1953.

——— and Kaylor, Robert. "Dishpan Hurricanes Help Study of Large Storms." *Science Newsletter,* Vol. 71:201, 1957.

Juisto, James E. and Pilie, Roland J. "Condensation Nuclei Experiments with Simple Apparatus." *Weatherwise,* December 1958, pp. 206–8.

Laird, Charles A. "How to Construct a Density Channel." *Weatherwise,* August 1948, p. 87.

———. "How to Construct a Simple Electrometer." *Weatherwise,* June 1949, p. 68.

Lettau, H. H. and Sparkman, J. K. "Hot Plate Mirages." Unpublished paper, Department of Meteorology, University of Wisconsin.

Lokke, Donald H. and Lokke, Virginia S. "Preparing Your Own Cloud Atlas." *Weatherwise,* August 1961, pp. 147–50.

Miller, James E. "A Tornado Model and the Fire Whirlwind." *Weatherwise,* Vol. 8, No. 3, p. 88.

Schaefer, Vincent J. "Preparation of Permanent Replicas of Snow, Frost, and Ice." *Weatherwise,* December 1964, pp. 279–87.

Stong, C. L. "The Amateur Scientist." *Scientific American,* March 1959, pp. 155–64.

———. "The Amateur Scientist." *Scientific American,* May 1963, pp. 167–74.

Storey, L. R. D. "Whistlers." *Scientific American,* January 1956, pp. 34–37.

Tanner, C. B. and Suomi, V. E. "Lithium Chloride Dew cell—Properties and Use for Dew-Point and Vapor Pressure Gradient Measurements." *Transactions, American Geophysical Union,* Vol. 37, No. 4, August 1956, pp. 413–20.

———. "A Max-Min Dew-Point Hygrometer." *Transactions, American Geophysical Union,* Vol. 39, No. 1, February 1958, pp. 63–66.

Ward, N. B. "Temperature Inversions as a Factor in Formation of Tornadoes." *Bulletin of American Meteorological Society,* Vol. 37, pp. 145–51.

Woodcock, A. H. "Salt and Rain." *Scientific American,* October 1957, pp. 42–47.

INDEX